Essential Calculus

Essential Calculus

Betty Reads

MURPHY & MOORE
www.murphy-moorepublishing.com

Published by Murphy & Moore Publishing,
1 Rockefeller Plaza,
New York City, NY 10020, USA

ISBN: 978-1-63987-199-5

Cataloging-in-Publication Data

Essential calculus / Betty Reads.
p. cm.
Includes bibliographical references and index.
ISBN 978-1-63987-199-5
1. Calculus. 2. Mathematical analysis. 3. Functions. 4. Geometry, Infinitesimal. I. Reads, Betty.
QA303.2 .E87 2022
515--dc23

For information on all Murphy & Moore Publications
visit our website at www.murphy-moorepublishing.com

MURPHY & MOORE

Contents

Preface

Calculus is the branch of mathematics which is concerned with the study of rates of change. It is further divided into two branches, namely, differential calculus and integral calculus. Differential calculus studies the rate of changes of functions with respect to their variables through the use of derivatives and differentials. Slopes and curves of graphs are a part of differential calculus. Integral calculus focuses on accumulation of quantities and studies the area under and between the curves. Calculus finds its applications across various domains such as robotics, video games, artificial intelligence, planetary motion, aerospace engineering, bridge engineering, business and economics. This book unfolds the innovative aspects of calculus which will be crucial for the holistic understanding of the subject matter. Some of the diverse topics covered herein address the varied branches that fall under this category. Those in search of information to further their knowledge will be greatly assisted by this book.

A detailed account of the significant topics covered in this book is provided below:

Chapter 1- The mathematical study which focuses on continuous change is called calculus. It deals with functions and limits. Some of the different types of functions used within calculus are implicit and explicit functions, and inverse functions. This is an introductory chapter which will provide a brief introduction to these aspects of calculus.

Chapter 2- The area of study which deals with the definition, properties and applications of the derivative of a function is called differential calculus. The process through which the derivative is found is termed as differentiation. The topics elaborated in this chapter will help in gaining a better perspective about differential calculus as well as the rules of differentiation.

Chapter 3- The process of differentiation can be used repeatedly, resulting in the second derivative of the function. It generally has a useful physical interpretation. This chapter closely examines the key concepts of second order derivatives and higher order derivatives as well as Leibniz formula to provide an extensive understanding of the subject.

Chapter 4- The study of definitions, properties and applications of definite integral and indefinite integral are studied within the field of integral calculus. The process through which the value of an integral is determined is called integration. All the diverse principles of integral calculus such as line integral and Reimann integral have been carefully analyzed in this chapter.

Chapter 5- There are various methods which are used for integration such as disc integration, shell integration, integration by substitution, trigonometric substitution, order of integration and integration by reduction. The topics elaborated in this chapter will help in gaining a better perspective about these methods of integration.

Chapter 6- The integration and differentiation of vector fields is termed as vector calculus. There are various theorems which are studied within this field such as Stokes' theorem, Green's theorem and divergence theorem. The topics elaborated in this chapter will help in gaining a better perspective about these aspects of vectors and vector calculus.

It gives me an immense pleasure to thank our entire team for their efforts. Finally in the end, I would like to thank my family and colleagues who have been a great source of inspiration and support.

Betty Reads

1

Understanding Calculus

The mathematical study which focuses on continuous change is called calculus. It deals with functions and limits. Some of the different types of functions used within calculus are implicit and explicit functions, and inverse functions. This is an introductory chapter which will provide a brief introduction to these aspects of calculus.

Calculus is a branch of mathematics that involves the study of rates of change. Before calculus was invented, math was static: It could only help calculate objects that were perfectly still. But the universe is constantly moving and changing. No objects—from the stars in space to subatomic particles or cells in the body—are always at rest. Indeed, just about everything in the universe is constantly moving. Calculus helped to determine how particles, stars, and matter actually move and change in real time.

Calculus is used in a multitude of fields that you wouldn't ordinarily think would make use of its concepts. Among them are physics, engineering, economics, statistics, and medicine. Calculus is also used in such disparate areas as space travel, as well as determining how medications interact with the body, and even how to build safer structures.

Calculus was developed in the latter half of the 17th century by two mathematicians, Gottfried Leibniz and Isaac Newton. Newton first developed calculus and applied it directly to the understanding of physical systems. Independently, Leibniz developed the notations used in calculus. Put simply, while basic math uses operations such as plus, minus, times, and division (+, -, x, and ÷), calculus uses operations that employ functions and integrals to calculate rates of change.

Those tools allowed Newton, Leibniz, and other mathematicians who followed to calculate things like the exact slope of a curve at any point. The Story of Mathematics explains the importance of Newton's fundamental theorem of the calculus: "Unlike the static geometry of the Greeks, calculus allowed mathematicians and engineers to make sense of the motion and dynamic change in the changing world around us, such as the orbits of planets, the motion of fluids, etc."

Using calculus, scientists, astronomers, physicists, mathematicians, and chemists could now chart the orbit of the planets and stars, as well as the path of electrons and protons at the atomic level.

Differential vs. Integral Calculus

There are two branches of calculus: Differential and integral calculus. Differential calculus studies the derivative and integral calculus studies the integral. But there is more to it than that. Differential calculus determines the rate of change of a quantity. It examines the rates of change of slopes and curves. This branch is concerned with the study of the rate of change of functions with respect to their variables, especially through the use of derivatives and differentials. The derivative is the slope of a line on a graph. You find the slope of a line by calculating the rise over the run.

Integral calculus, by contrast, seeks to find the quantity where the rate of change is known. This branch focuses on such concepts as slopes of tangent lines and velocities. While differential calculus focuses on the curve itself, integral calculus concerns itself with the space or area under the curve. Integral calculus is used to figure the total size or value, such as lengths, areas, and volumes.

Calculus played an integral role in the development of navigation in the 17th and 18th centuries because it allowed sailors to use the position of the moon to accurately determine the local time. To chart their position at sea, navigators needed to be able to measure both time and angles with accuracy. Before the development of calculus, ship navigators and captains could do neither.

Calculus — both derivative and integral — helped to improve the understanding of this important concept in terms of the curve of the Earth, the distance ships had to travel around a curve to get to a specific location, and even the alignment of the Earth, seas, and ships in relation to the stars.

Practical Applications

Calculus has many practical applications in real life. Some of the concepts that use calculus include motion, electricity, heat, light, harmonics, acoustics, and astronomy. Calculus is used in geography, computer vision (such as for autonomous driving of cars), photography, artificial intelligence, robotics, video games, and even movies. Calculus is also used to calculate the rates of radioactive decay in chemistry, and even to predict birth and death rates, as well as in the study of gravity and planetary motion, fluid flow, ship design, geometric curves, and bridge engineering.

In physics, for example, calculus is used to help define, explain, and calculate motion, electricity, heat, light, harmonics, acoustics, astronomy, and dynamics. Einstein's theory of relativity relies on calculus, a field of mathematics that also helps economists predict how much profit a company or industry can make. And in shipbuilding, calculus has been used for many years to determine both the curve of the hull of the ship (using differential calculus), as well as the area under the hull (using integral calculus), and even in the general design of ships. In addition, calculus is used to check answers for different mathematical disciplines such as statistics, analytical geometry, and algebra.

Function and its Types

A function is a relation between a set of inputs and a set of permissible outputs with the property that each input is related to exactly one output. Let A & B be any two non-empty sets, mapping from A to B will be a function only when every element in set A has one end only one image in set B.

Example:

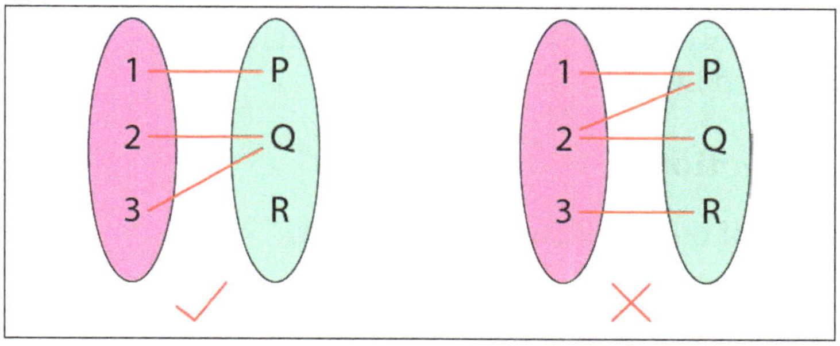

Another definition of functions is that it is a relation "f" in which each element of set "A" is mapped with only one element belonging to set "B". Also in a function, there can't be two pairs with the same first element.

A Condition for a Function

Set A and Set B should be non-empty. In a function, a particular input is given to get a particular output. So, A function f: A->B denotes that f is a function from A to B, where A is a domain and B is a co-domain. For an element, a, which belongs to A, a∈A, a unique element b, b∈B is there such that (a,b) ∈ f.

The unique element b to which f relates a, is denoted by f(a) and is called f of a, or the value of f at a, or the image of a under f.

- The range of f (image of a under f).
- It is the set of all values of f(x) taken together.
- Range of f = {y ∈ Y|y = f (x), for some x in X}.

A real-valued function has either P or any one of its subsets as its range. Further, if its domain is also either P or a subset of P, it is called a real function.

Vertical Line Test

Vertical line test is used to determine whether a curve is a function or not. If any curve cuts a vertical line at more than one point then the curve is not a function.

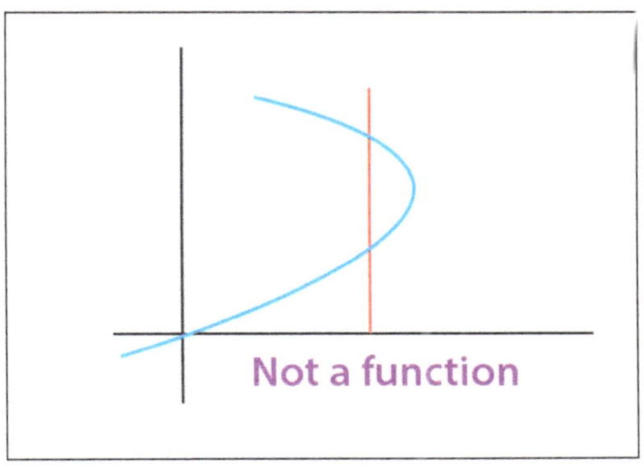

Representation of Functions

Functions are generally represented as f(x). Let, $f(x) = x^3$. It is said as f of x is equal to x cube. Functions can also be represented by g(), t(),... etc.

Steps for Solving Functions

Question: Find the output of the function $g(t) = 6t^2 + 5$ at,

- $t = 0$
- $t = 2$

Solution: The given function is $g(t)= 6t^2 + 5$,

- At $t = 0$, $g(0) = 6(0)^2 + 5 = 5$
- At $t = 2$, $g(2) = 6(2)^2 + 5 = 29$

Types of Functions

There are various types of functions in mathematics which are:

- One: One function (Injective function),
- Many: One function,
- Onto Function (Surjective Function),
- Into Function,
- Polynomial Function,
- Linear Function,
- Identical Function,
- Quadratic Function,
- Rational Function,
- Algebraic Functions,
- Cubic Function,
- Modulus Function,
- Signum Function,
- Greatest Integer Function,
- Fractional Part Function,
- Even and Odd Function,

- Periodic Function,
- Composite Function,
- Constant Function,
- Identity Function.

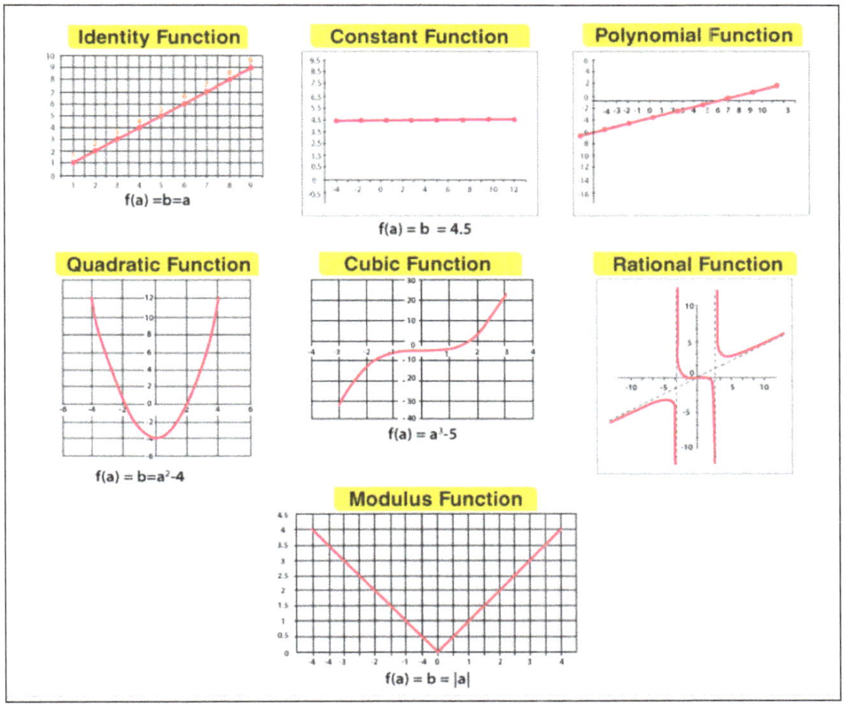

One – one Function (Injective Function)

If each element in the domain of a function has a distinct image in the co-domain, the function is said to be one – one function. For examples f; R R given by f(x) = 3x + 5 is one – one.

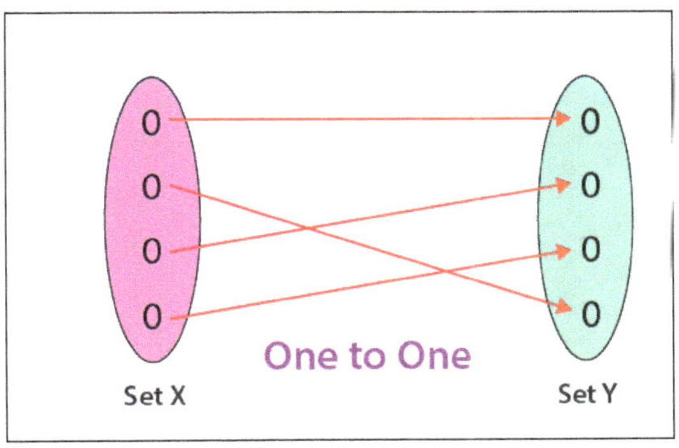

Many – one Function

On the other hand, if there are at least two elements in the domain whose images are same, the function is known as many to one. For example f: R R given by f(x) = x^2 + 1 is many one.

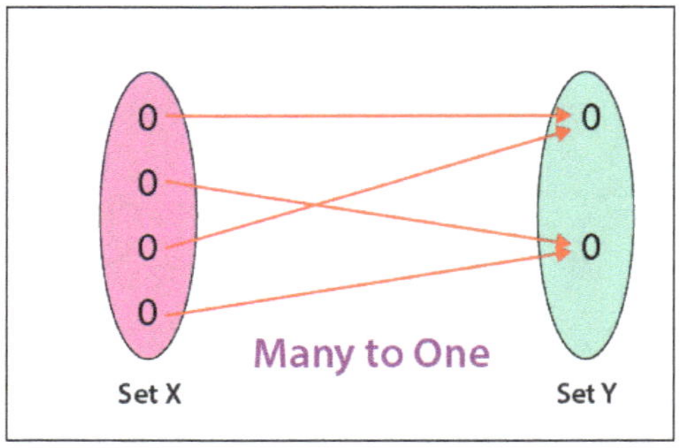

Onto – function (Surjective Function)

A function is called an onto function if each element in the co-domain has at least one pre – image in the domain.

Into – function

If there exists at least one element in the co-domain which is not an image of any element in the domain then the function will be into function. (Q) Let A = {x : 1 < x < 1} = B be a mapping f : A B, find the nature of the given function (P). F(x) = |x|, f (x) = |1|.

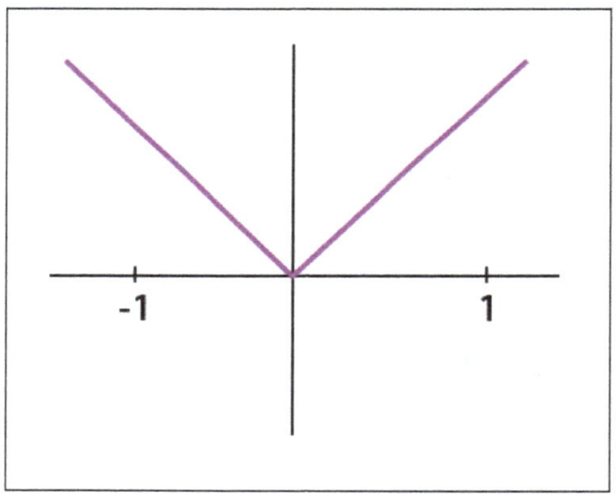

Solution for x = 1 & -1. Hence it is many one the Range of f(x) from [-1, 1] is [0,1] which is not equal to co-domain. Hence it is into function.

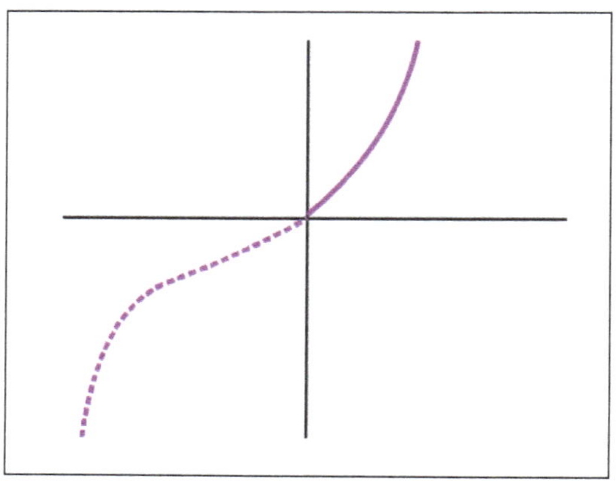

Let's say we have function,

$$f(x)=\begin{cases} x^2 & ; x \geq 0 \\ -x^2 & ; x < 0 \end{cases}$$

For different values of Input, we have different output hence it is one – one function also it manage is equal to its co-domain hence it is onto also.

Polynomial Function

A real valued function f : P → P defined by $y=f(a)=h_0+h_1a+......+h_na^n$, where n ϵ N, and $h_0+h_1+.....+h_n \in P$, for each a ϵ P, is called polynomial function.

- N = a non-negative integer.
- The degree of Polynomial function is the highest power in the expression.
- If the degree is zero, it's called a constant function.
- If the degree is one, it's called a linear function. Example: b = a+1.
- Graph type: Always a straight line.

So, a polynomial function can be expressed as:

$$f(x)=a_nx^n+a_{n-1}x^{n-1}+.....+a_1x^1+a_0$$

The highest power in the expression is known as the degree of the polynomial function. The different types of polynomial functions based on the degree are:

- The polynomial function is called a Constant function if the degree is zero.
- The polynomial function is called a Linear if the degree is one.
- The polynomial function is Quadratic if the degree is two.
- The polynomial function is Cubic if the degree is three.

Linear Function

All functions in the form of ax + b where a, b ∈ R & a ≠ 0 are called as linear functions. The graph will be a straight line. In other words, a linear polynomial function is a first-degree polynomial where the input needs to be multiplied by m and added to c. It can be expressed by f(x) = mx + c.

For example:

- f(x) = 2x + 1 at x = 1
- f(1) = 2.1 + 1 = 3
- f(1) = 3

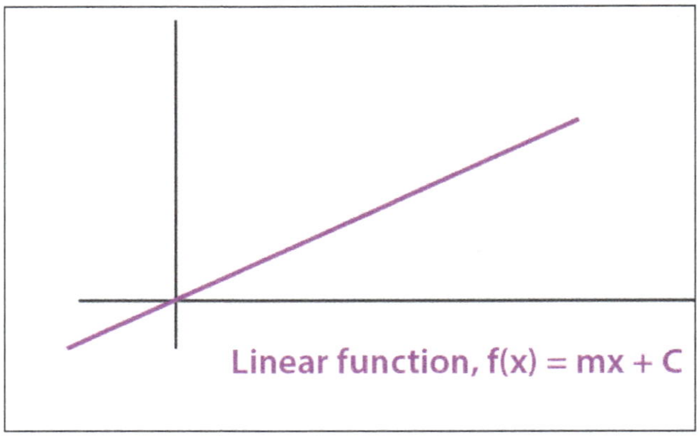

Another example of linear function is y = x + 3.

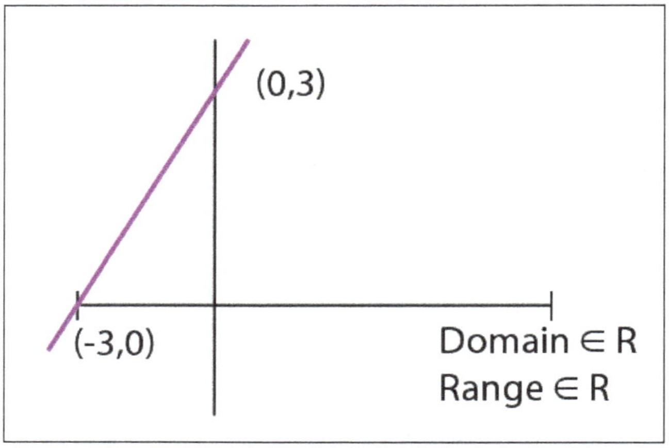

Identical Function

Two functions f and g are said to be identical if:

- The domain of f = domain of g.
- The range of f = the Range of g.
- $f(x)=g(x)\forall x\in D_f$ & D_g.

For example, $f(x)=x$ & $g(x)=\frac{1}{1/x}$.

Solution: f(x) = x is defined for all x. But $g(x)=\frac{1}{1/x}$ is not defined of x = 0. Hence it is identical for x ∈ R {0}.

Quadratic Function

All functions in the form of $y = ax^2 + bx + c$ where a, b, c ∈ R, a ≠ 0 will be known as Quadratic function. The graph will be parabolic.

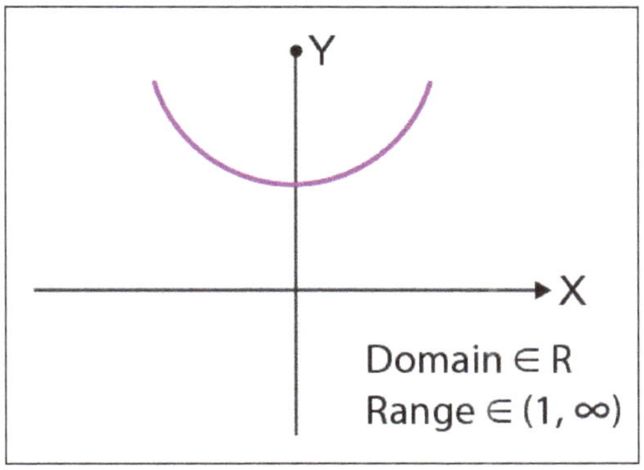

At $x = \frac{-b \pm \sqrt{D}}{2}$, we will get its maximum on minimum value depends on the leading coefficient and that value will be $\frac{-D}{4a}$ (where D = Discriminant).

In simpler terms, A Quadratic polynomial function is a second degree polynomial and it can be expressed as, $F(x) = ax^2 + bx + c$, and a is not equal to zero. Where a, b, c are constant and x is a variable.

Example: $f(x) = 2x^2 + x - 1$ at $x = 2$. If $x = 2$, $f(2) = 2.2^2 + 2 - 1 = 9$.

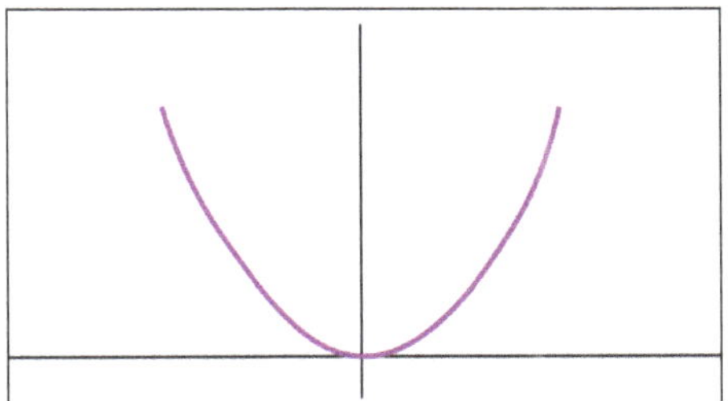

Quadratic function, $f(x) = x^2$

For example: $y = x^2 + 1$.

Rational Function

These are the real functions of the type $\frac{f(a)}{g(a)}$ where f(a) and g(a) are polynomial functions of a defined in a domain, where $g(a) \neq 0$.

- For example: $f : P - \{-6\} \rightarrow P$ defined by $f(a) = \frac{f(a+1)}{g(a+2)}$, $\forall a \epsilon P - \{-6\}$ is a rational function.
- Graph type: Asymptotes (the curves touching the axes lines).

Algebraic Functions

A function that consists of a finite number of terms involving powers and roots of independent variable x and fundamental operations such as addition, subtraction, multiplication, and division is known as an algebraic equation.

For example,

$$f(x) = 5x^3 - 2x^2 + 3x + 6,\ g(x) = \frac{\sqrt{3x+4}}{(x-1)^2}.$$

Cubic Function

A cubic polynomial function is a polynomial of degree three and can be expressed as; $F(x) = ax^3 + bx^2 + cx + d$ and a is not equal to zero.

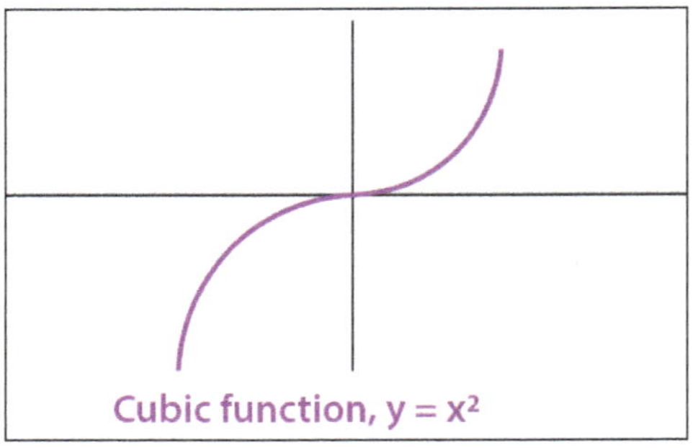

In other words, any function in the form of $f(x) = ax^3 + bx^2 + cx + d$, where $a, b, c, d \in R$ & $a \neq 0$.

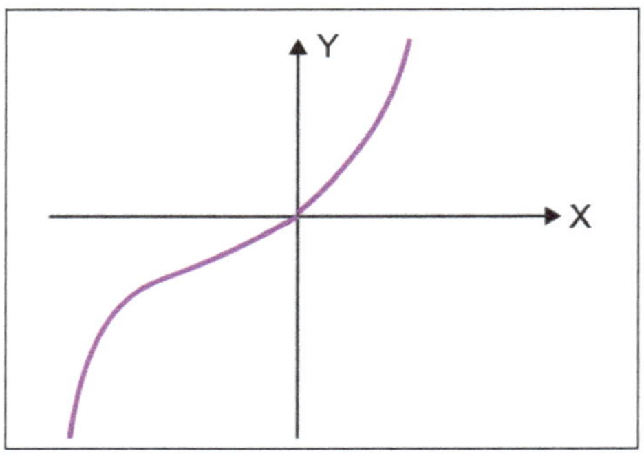

For example: $y = x^3$

Domain $\in$ R

Range $\in$ R

Modulus Function

The real function $f : P \to P$ defined by $f(a) = f(a) = |a| = a$, $a \geq 0$, and $f(a) = -a$ if $a<0$, $\forall a \in P$ is called the modulus function.

- Domain of f = P
- Range of f = $P^+ \cup 0$

$$y = |x| = \begin{cases} x & x \geq 0 \\ -x & x < 0 \end{cases}$$

Domain: R.

Range: [0, ∞).

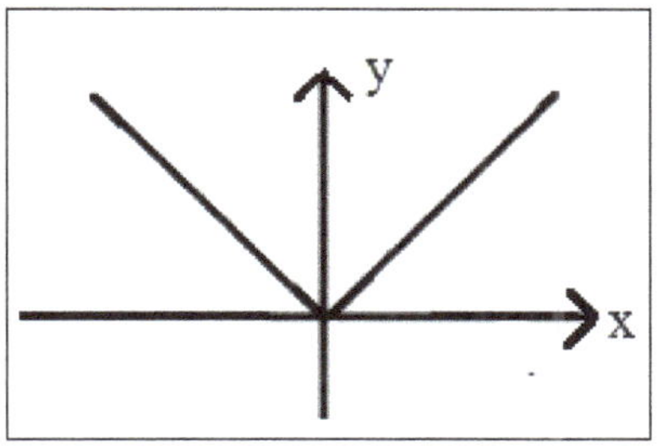

Signum Function

The real function f : P → P defined by,

$$\begin{cases} \frac{|f(a)|}{f(a)}, & a \neq 0 \\ 0, & a = 0 \end{cases} = \begin{cases} 1, \text{if } x > 0 \\ 0, \text{if } x = 0 \\ -1, \text{ if } x < 0 \end{cases}$$

is called the signum function or sign function. (Gives the sign of real number).

- Domain of f = P.
- Range of f = {1, 0, – 1}.

For example:

- Signum (100) = 1.
- Signum (log 1) = 0.
- Signum ($x^2$1) =1.

Greatest Integer Function

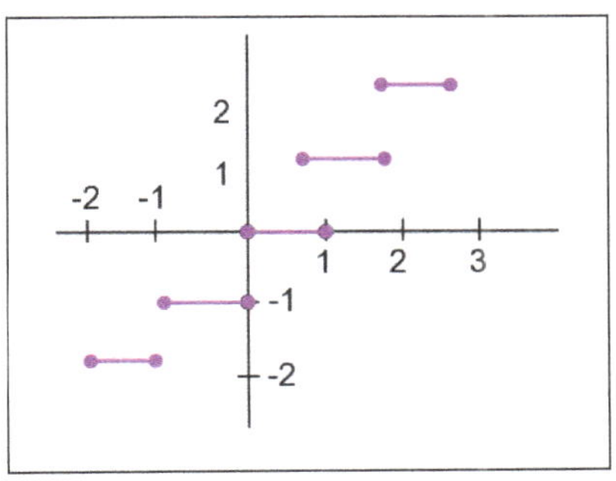

The real function f : P → P defined by f (a) = [a], a ϵ P assumes the value of the greatest integer less than or equal to a, is called the greatest integer function.

- Thus f (a) = [a] = – 1 for – 1 ⩽ a < 0.

- f (a) = [a] = 0 for 0 ⩽ a < 1.
- [a] = 1 for 1 ⩽ a < 2.
- [a] = 2 for 2 ⩽ a < 3 and so on.

The greatest integer function always gives integral output. The Greatest integral value that has been taken by the input will be the output.

For example:

- [4.5] = 4
- [6.99] = 6
- [1.2] = 2
- Domain ∈R
- Range ∈ Integers

Fractional Part Function

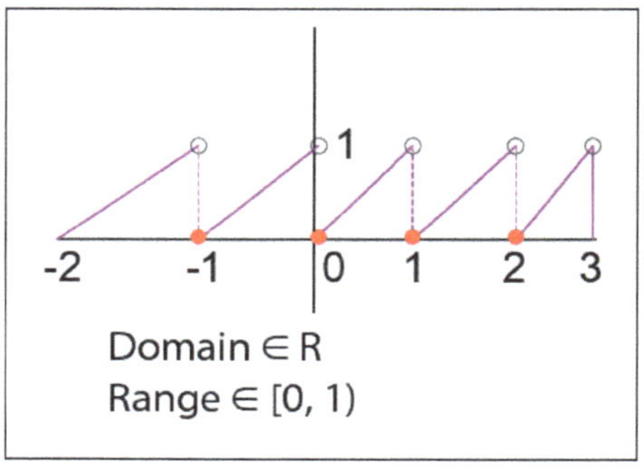

$$\{x\} = x - [x]$$

It always give fractional value as output.

For example:

- {4.5} = 4.5 – [4.5] = 4.5 – 4 = 0.5
- {6.99} = 6.99 – [6.99] = 6.99 – 6 = 0.99
- {7} = 7 – [7] = 7 –7 = 0

Even and Odd Function

If f(x) = f(-x) then the function will be even function & f(x) = -f(-x) then the function will be odd function. Example:

$$f(x) = x^2 \sin x$$
$$f(-x) = -x^2 \sin x$$

Here, f(x) = -f(-x) it is odd function.

Example: $f(x)=x^2$ and $f(-x)=x^2$ $f(x)=f(-x)$. It is an even function.

Periodic Function

A function is said to be a periodic function if there exist a positive real numbers T such that f(u – t) = f(x) for all x ε Domain. For example f(x) = sinx,

$$f(x+2\pi)=\sin\ (x+2\pi)=\sin x \text{ fundamental}$$

then period of sinx is 2π.

Composite Function

Let A, B, C be three non-empty sets.

Let f: A B & g : G C be two functions then gof : A C. This function is called composition of f and g given g of (x) = g(f(x)).

For example: f(x) = x² & g(x) = 2x

$$f(g(x))=f(2x)=(2x)^2=4x^2$$

$$g(f(x))=g(x^2)=2x^2$$

Constant Function

The function f : P → P defined by b = f (x) = D, a ϵ P, where D is a constant ϵ P, is a constant function.

- Domain of f = P.
- Range of f = {D}.
- Graph type: A straight line which is parallel to the x-axis.

In simple words, the polynomial of 0th degree where f(x) = f(0) = a_0 = c. Regardless of the input, the output always results in constant value. The graph for this is a horizontal line.

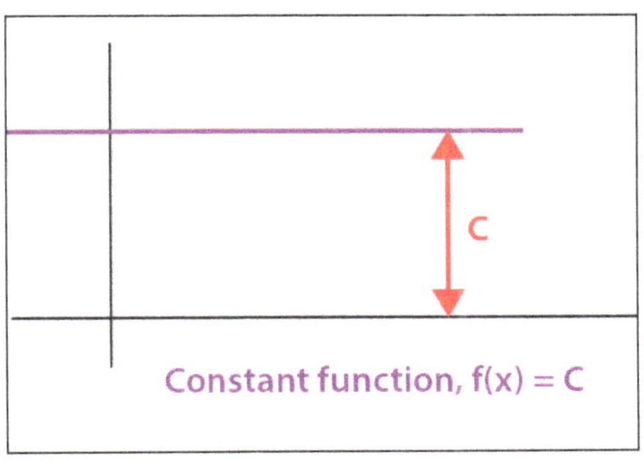

Identity Function

P = set of real numbers. The function f : P → P defined by b = f (a) = a for each a ϵ P is called the identity function.

- Domain of f = P.
- Range of f = P.
- Graph type: A straight line passing through the origin.

Implicit and Explicit Function

The explicit function is a function in which the dependent variable has been given "explicitly" in terms of the independent variable. Or it is a function in which the dependent variable is expressed in terms of some independent variables. It is denoted by: y=f(x).

Examples of Explicit functions are:

- $y = ax^n + bx$ where a, n and b are constant.
- $y = 5x^3-3$.

The Implicit function is a function in which the dependent variable has not been given "explicitly" in terms of the independent variable. Or it is a function in which the dependent variable is not expressed in terms of some independent variables. It is denoted by: R(x,y) = 0.

Some examples of Implicit Functions are:

$$x^2 + y^2 - 1 = 0$$
$$y^4 + x^3 + 17 = 0$$

Although you can convert an Implicit function into Explicit function it is generally not done because after conversion the new explicit function becomes very complex and sometimes also gives two different function branch. For example you can convert the first implicit function example above to an Explicit function but after conversion it gives the following function:

$$y = \pm\sqrt{1-x^2}$$

And in the new function there are two branches of "y" one the positive branch and another the negative branch.

Inverse functions

An inverse function is a function that will "undo" anything that the original function does. For example, we all have a way of tying our shoes, and how we tie our shoes could be called a function. So, what would be the inverse function of tying our shoes? The inverse function would be "untying" our shoes, because "untying" our shoes will "undo" the original function of tying our shoes.

Let's look at an inverse function from a mathematical point of view. Consider the function f(x) = 2x – 5. If we take any value of x and plug it into f(x) what would happen to that value of x? First, the value of x would get multiplied by 2 and then we would subtract 5. The two mathematical operations that are taking place in the function f(x) are multiplication and subtraction. Now let's consider the inverse function. What two mathematical operations would be needed to "undo" f(x)? Division and addition would be needed to "undo" the multiplication and subtraction. We will find the inverse of $f(x) = 2x - 5$, and hopefully the inverse function will contain both division and addition.

Notation

If f(x) represents a function, then the notation $f^{-1}(x)$, read "f inverse of x", will be used to denote the inverse of f(x). Similarly, the notation $g^{-1}(x)$, read "g inverse of x", will be used to denote the the inverse of g(x).

$f^{-1}(x) \neq \frac{1}{f(x)}$ · It is very important not to confuse function notation with negative exponents.

Does the Function have an Inverse?

Not all functions have an inverse, so it is important to determine whether or not a function has an inverse before we try and find the inverse. If a function does not have an inverse, then we need to realize the function does not have an inverse so we do not waste time trying to find something that does not exist.

So how do we know if a function has an inverse? To determine if a function has an inverse function, we need to talk about a special type of function called a One to One Function. A one to one function is a function where each input (x value) has a unique output (y value). To put it another way, every time we plug in a value of x we will get a unique value of y, the same y value will never appear more than once. A one to one function is special because only one to one functions have an inverse function. Only One to One Functions have an inverse function.

Example: Determine if the function f = {(7, 3), (8, -5), (-2, 11), (-6, 4)} is a one to one function. The function f is a one to one function because each of the y values in the ordered pairs is unique; none of the y values appear more than once. Since the function f is a one to one function, the function f must have an inverse.

Example: Determine if the function h = {(-3, 8), (-11, −9), (5, 4), (6, -9)} is a one to one function. The function h is not a one to one function because the y value of -9 is not unique; the y value of -9 appears more than once. Since the function h is not a one to one function, the function h does not have an inverse. Remember that only one to one function have an inverse.

Is the Function a One to One Function?

We can determine if functions are one to one by looking at ordered pairs and determining if each of the y values is unique, but what if we do not have ordered pairs? We could create ordered pairs by plugging different x values into the function and finding the corresponding y values giving us some ordered pairs. Rather than spending time creating ordered pairs, why not consider looking at the entire graph of the function instead? By looking at the entire graph rather than a few points, we should still be able to determine if the function is a one to one function or not.

In looking at the graph of the function we can determine if a function is a one to one function or not by applying the Horizontal Line Test, or HLT. If the graph of the function passes the Horizontal Line Test, then the function is a one to one function. If the graph of the function fails the Horizontal Line Test, then the function is not a one to one function.

Horizontal Line Test: The HLT says that a function is a one to one function if there is no horizontal line that intersects the graph of the function at more than one point.

By applying the Horizontal Line Test not only can we determine if a function is a one to one function, but more importantly we can determine if a function has an inverse or not.

Example: Determine if the function $f(x) = -\frac{3}{4}x + 2$ is a one to one function. To determine if f(x) is a one to one function, we need to look at the graph of f(x). Since f(x) is a linear equation the graph of f(x) is a line with a slope of −3/4 and a y intercept of (0, 2).

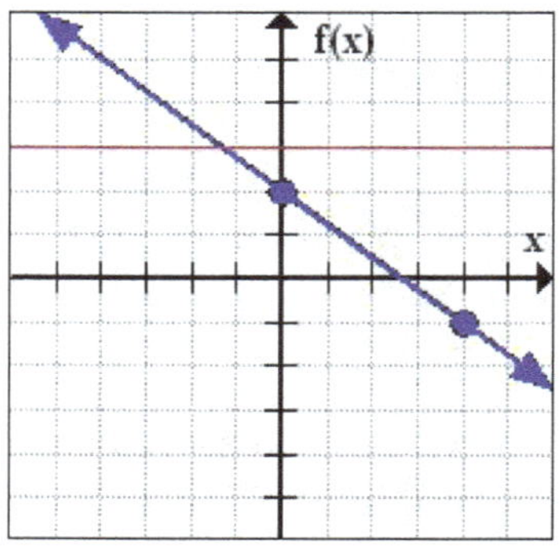

In looking at the graph, you can see that any horizontal line (shown in red) drawn on the graph will intersect the graph of f(x) only once. Therefore, f(x) is a one to one function and f(x) must have an inverse.

Example: Determine if the function $g(x) = x^3 - 4x$ is a one-to-one function. To determine if g(x) is a one-to-one function, we need to look at the graph of g(x).

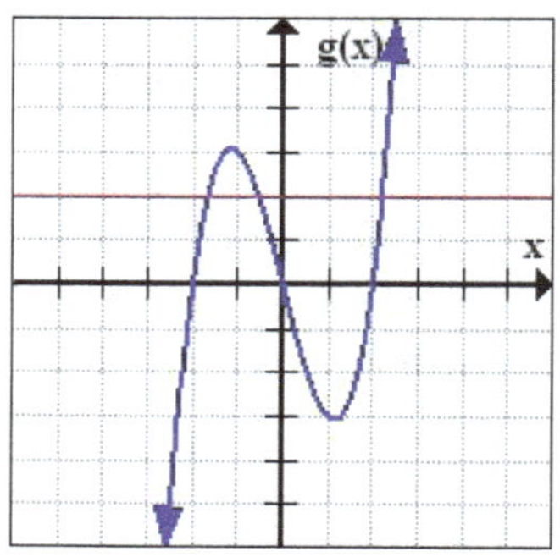

In looking at the graph, you can see that the horizontal line (shown in red) drawn on the graph intersects the graph of g(x) more than once. Therefore, g(x) is not a one to one function and g(x) does not have an inverse.

How to Find the Inverse Function

Now that we have discussed what an inverse function is, the notation used to represent inverse functions, one to one functions, and the Horizontal Line Test, we are ready to try and find an inverse function. Here are the steps required to find the inverse function:

- Step 1: Determine if the function has an inverse. Is the function a one to one function? If the function is a one to one function, go to step 2. If the function is not a one to one function, then say that the function does not have an inverse and stop.
- Step 2: Change f(x) to y.
- Step 3: Switch x and y.
- Step 4: Solve for y.
- Step 5: Change y back to $f^{-1}(x)$.

By following these 5 steps we can find the inverse function. Make sure that you follow all 5 steps. Many people will skip step 1 and just assume that the function has an inverse; however, not every function has an inverse, because not every function is a one to one function. Only functions that pass the Horizontal Line Test are one to one functions and only one to one functions have an inverse.

Example: If f(x) = 2x – 5, find the inverse.

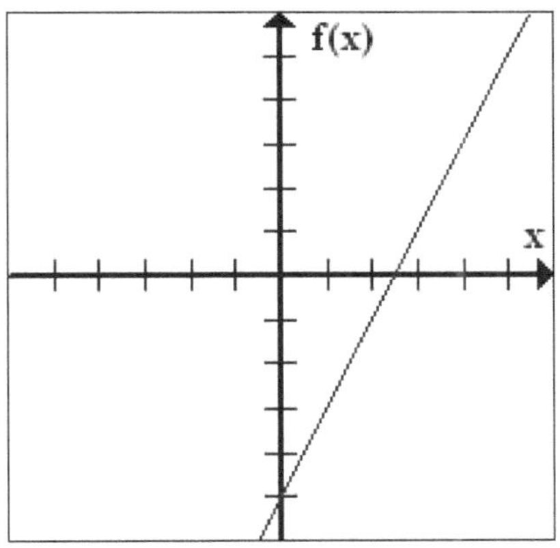

This function passes the Horizontal Line Test which means it is a one to one function that has an inverse.

$y = 2x - 5$	Change f(x) to y.
$x = 2y - 5$	Switch x and y.
$x + 5 = 2y$ $\frac{x+5}{2} = y$	Solve for y by adding 5 to each side and then dividing each side by 2.
$f^{-1}(x) = \frac{x+5}{2} = \frac{1}{2}x + \frac{5}{2}$	Change y back to f –1(x).

Therefore, $f^{-1}(x) = \frac{x+5}{2}$ or $f^{-1}(x) = \frac{x+5}{2} = \frac{1}{2}x + \frac{5}{2}$.

Trigonometric Functions

1. Trigonometric functions are elementary functions, the argument of which is an angle. Trigonometric functions describe the relation between the sides and angles of a right triangle. Applications of trigonometric functions are extremely diverse. For example, any periodic processes can be represented as a sum of trigonometric functions (Fourier series). These functions often appear in the solution of differential equations and functional equations.

2. The trigonometric functions include the following 6 functions: Sine, cosine, tangent, cotangent, secant, and cosecant. For each of these functions, there is an inverse trigonometric function.

3. The trigonometric functions can be defined using the unit circle. The figure below shows a circle of radius r = 1. There is a point M(x,y) on the circle. The angle between the radius vector OM and the positive direction of the x-axis is equal to α.

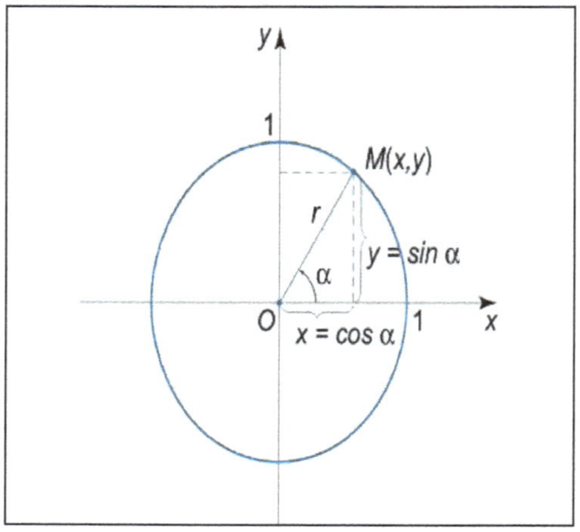

4. The sine of an angle α is the ratio of the y-coordinate of the point M(x,y) to the radius r: sin α=y/r. Since r=1, the sine is equal to the y-coordinate of the point M(x,y).

5. The cosine of an angle α is the ratio of the x-coordinate of the point M(x,y) to the radius r: cos α = x/r.

6. The tangent of an angle α is the ratio of the y-coordinate of the point M(x,y) to the x-coordinate: tanα = y/x, x $\neq$ 0.

7. The cotangent of an angle α is the ratio of the x-coordinate of the point M(x,y) to the y-coordinate: cot α = x/y, y $\neq$ 0.

8. The secant of an angle α is the ratio of the radius r to the x-coordinate of the point M(x,y): sec α = r/x = 1/x, x $\neq$ 0.

9. The cosecant of an angle α is the ratio of the radius r to the y-coordinate of the point M(x,y): csc α = r/y = 1/y, y $\neq$ 0.

10. Relation between the sides and angles in a right triangle. In a unit circle, the projections x, y of

a point M(x,y) and the radius r form a right-angled triangle, in which x,y are the legs, and r is the hypotenuse. Therefore, the definitions given above are stated as follows:

- The sine of an angle α is the ratio of the opposite leg to the hypotenuse.
- The cosine of an angle α is the ratio of the adjacent leg to the hypotenuse.
- The tangent of an angle α is the ratio of the opposite leg to the adjacent leg.
- The cotangent of an angle α is the ratio of the adjacent leg to the opposite leg.
- The secant of an angle α is the ratio of the hypotenuse to the adjacent leg.
- The cosecant of an angle α is the ratio of the hypotenuse to the opposite leg.

11. Graph of the sine function: $y = \sin x$, domain: $x \in \mathbb{R}$, range: $-1 \le \sin x \le 1$.

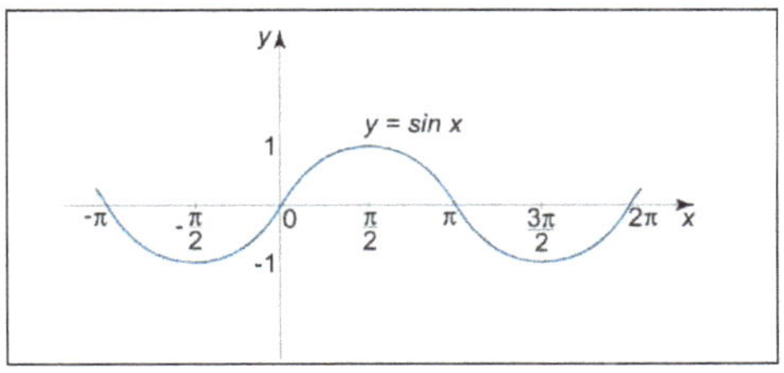

12. Graph of the cosine function: $y = \cos x$, domain: $x \in \mathbb{R}$, range: $-1 \le \cos x \le 1$.

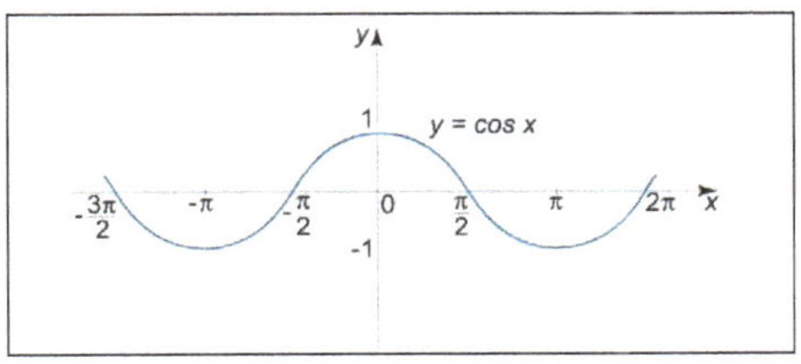

13. Graph of the tangent function: $y = \tan x$, domain: $x \in \mathbb{R}$, $x \neq (2k+1)\pi/2$, range: $-\infty < \tan x < \infty$.

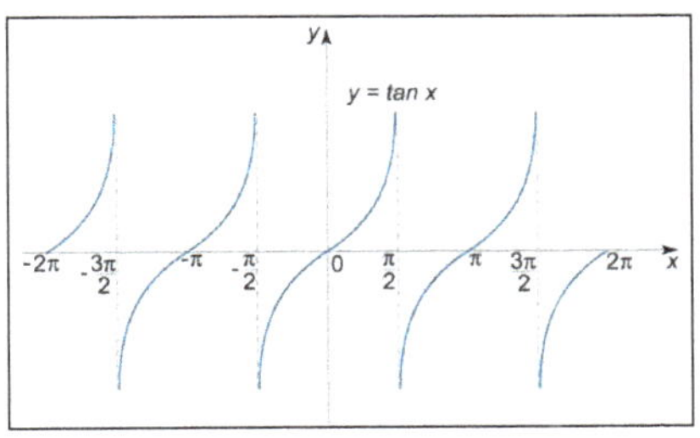

14. Graph of the cotangent function: $y = \cot x$, domain: $x \in \mathbb{R}$, $x \neq k\pi$, range: $-\infty < \cot x < \infty$.

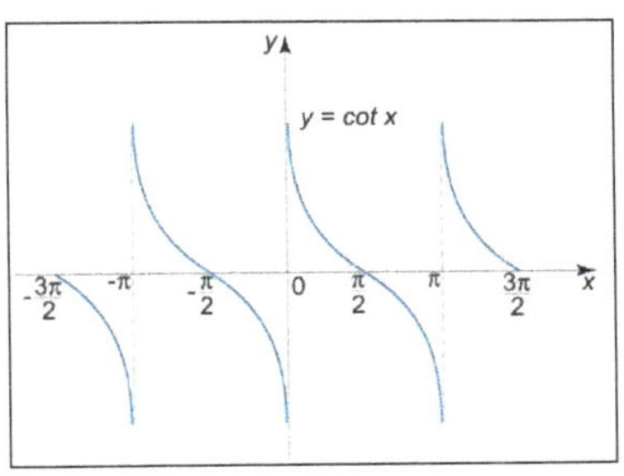

15. Graph of the secant function: $y = \sec x$, domain: $x \in \mathbb{R}, x \neq (2k+1)\pi/2$, range: $\sec x \in (-\infty, -1] \cup [1,\infty)$.

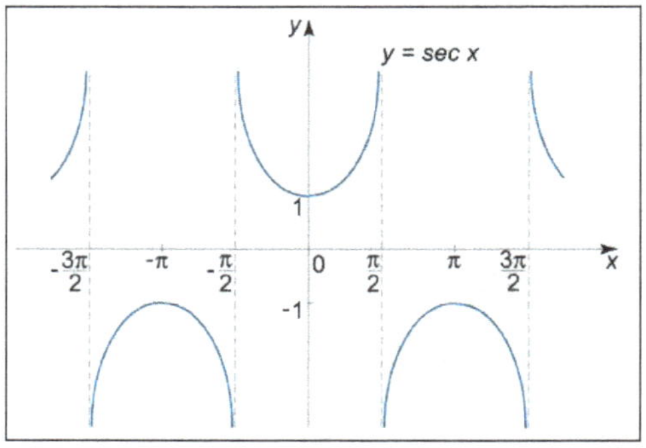

16. Graph of the cosecant function: $y = \csc x$, domain: $x \in \mathbb{R}, x \neq k\pi$, range: $\csc x \in (-\infty,-1] \cup [1,\infty)$.

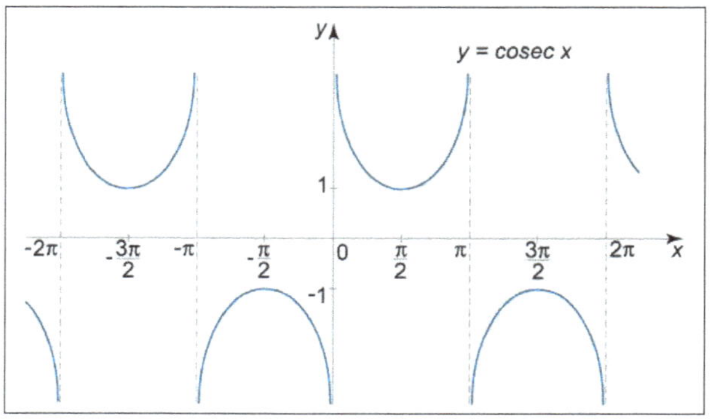

Functions in Several Variables

The definition of a function of two variables is very similar to the definition for a function of one variable. The main difference is that, instead of mapping values of one variable to values of another variable, we map ordered pairs of variables to another variable.

A function of two variables $z = (x,y)$ maps each ordered pair (x,y) in a subset D of the real plane IR^2 to a unique real number z. The set D is called the domain of the function. The range of f is the set of all real numbers z that has at least one ordered pair $(x,y) \in D$ such that $f(x,y) = z$ as shown in figure.

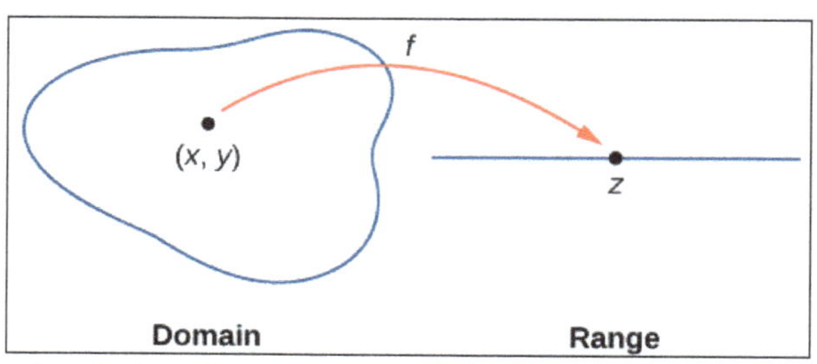

The domain of a function of two variables consists of ordered pairs (x, y).

Determining the domain of a function of two variables involves taking into account any domain restrictions that may exist.

Example: Domains and ranges for functions of two variables. Find the domain and range of each of the following functions:

- $f(x,y)=3x+5y+2$
- $g(x,y)=\sqrt{9-x^2-y^2}$

Solution: This is an example of a linear function in two variables. There are no values or combinations of x and y that cause f(x,y) to be undefined, so the domain of f is $\mathbb{R}^2$. Written in set-builder notation, this could be written as, $\{(x,y)\,|\,x\in\mathbb{R}, y\in\mathbb{R}\}$.

To determine the range, first pick a value for z. We need to find a solution to the equation f(x,y) = z, or 3x − 5y + 2 = z. One such solution can be obtained by first setting y = 0, which yields the equation 3x + 2 = z. The solution to this equation is $x=\frac{z-2}{3}$, which gives the ordered pair $\left(\frac{z-2}{3},0\right)$ as a solution to the equation f(x,y) = z for any value of z. Therefore, the range of the function is all real numbers, or $(-\infty,\infty)$.

For the function g(x,y) to have a real value, the quantity under the square root must be nonnegative:

$$9-x^2-y^2\geq 0.$$

This inequality can be written in the form,

$$x^2+y^2\leq 9.$$

Therefore, the domain of g(x,y) is $\{(x,y)\in R^2\,|\,x^2+y^2\leq 9\}$. The graph of this set of points can be described as a disk of radius 3 centered at the origin. The domain includes the boundary circle as shown in the following graph:

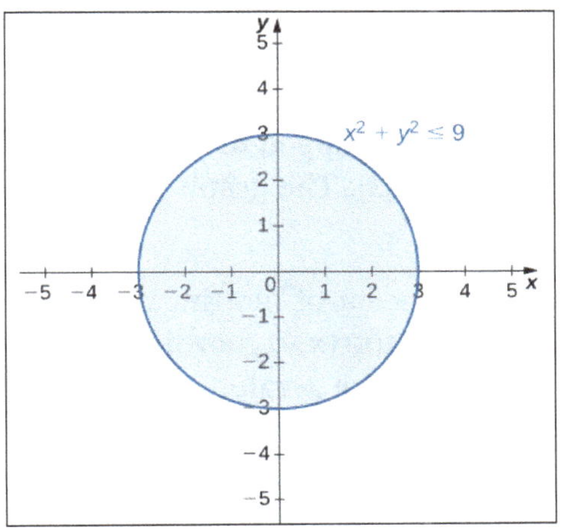

The domain of the function $g(x,y)=\sqrt{9-x^2-y^2}$ is a closed disk of radius 3.

To determine the range of $g(x,y)=\sqrt{9-x^2-y^2}$ we start with a point (x_0, y_0) on the boundary of the domain, which is defined by the relation $x^2 + y^2 = 9$. It follows that $x_0^2+y_0^2=9$ and,

$$\begin{aligned} g(x_0,y_0) &= \sqrt{9-x_0^2-y_0^2} \\ &= \sqrt{9-(x_0^2-y_0^2)} \\ &= \sqrt{9-9} \\ &= 0. \end{aligned}$$

If $x_0^2+y_0^2=0$ (in other words, $x_0 = y_0 = 0$), then,

$$\begin{aligned} g(x_0,y_0) &= \sqrt{9-x_0^2-y_0^2} \\ &= \sqrt{9-(x_0^2+y_0^2)} \\ &= \sqrt{9-0}=3. \end{aligned}$$

This is the maximum value of the function. Given any value c between 0 and 3, we can find an entire set of points inside the domain of g such that g(x,y) = c:

$$\begin{aligned} \sqrt{9-x^2-y^2} &= c \\ 9-x^2-y^2 &= c^2 \\ x^2+y^2 &= 9-c^2. \end{aligned}$$

Since $9 - c^2 > 0$, this describes a circle of radius $\sqrt{9-c^2}$ centered at the origin. Any point on this circle satisfies the equation g(x,y) = c. Therefore, the range of this function can be written in interval notation as [0,3].

Graphing Functions of Two Variables

Suppose we wish to graph the function z = (x,y). This function has two independent variables (x and y) and one dependent variable (z). When graphing a function y = f(x) of one variable, we use the Cartesian plane. We are able to graph any ordered pair (x,y) in the plane, and every point in the plane has an ordered pair (x,y) associated with it. With a function of two variables, each ordered pair (x,y) in the domain of the function is mapped to a real number z. Therefore, the graph of the function f consists of ordered triples (x,y,z). The graph of a function z = (x,y) of two variables is called a surface.

To understand more completely the concept of plotting a set of ordered triples to obtain a surface in three-dimensional space, imagine the (x,y) coordinate system laying flat. Then, every point in the domain of the function f has a unique z-value associated with it. If z is positive, then the graphed point is located above the xy-plane, if z is negative, then the graphed point is located below the xy-plane. The set of all the graphed points becomes the two-dimensional surface that is the graph of the function f.

Example: Graphing functions of two variables. Create a graph of each of the following functions:

- $g(x,y)=\sqrt{9-x^2-y^2}$

- $f(x,y)=x^2+y^2$

Solution: Here, we determined that the domain of $g(x,y)=\sqrt{9-x^2-y^2}$ is $\{(x,y)\in R^2 \mid x^2+y^2\le 9\}$ and the range is $\{z\in R^2 \mid 0\le z\le 3\}$. When $x^2 + y^2 = 9$ we have g(x,y) = 0. Therefore any point on the circle of radius 3 centered at the origin in the xy-plane maps to z = 0 in R^3. If $x^2 + y^2 = 8$, then g(x,y) = 1, so any point on the circle of radius $2\sqrt{2}$ centered at the origin in the xy-plane maps to z = 1 in R^3. As $x^2 + y^2$ gets closer to zero, the value of z approaches 3. When $x^2 + y^2 = 0$, then g(x,y) = 3. This is the origin in the xy-plane If $x^2 + y^2$ is equal to any other value between 0 and 9, then g(x,y) equals some other constant between 0 and 3. The surface described by this function is a hemisphere centered at the origin with radius 3 as shown in the following graph:

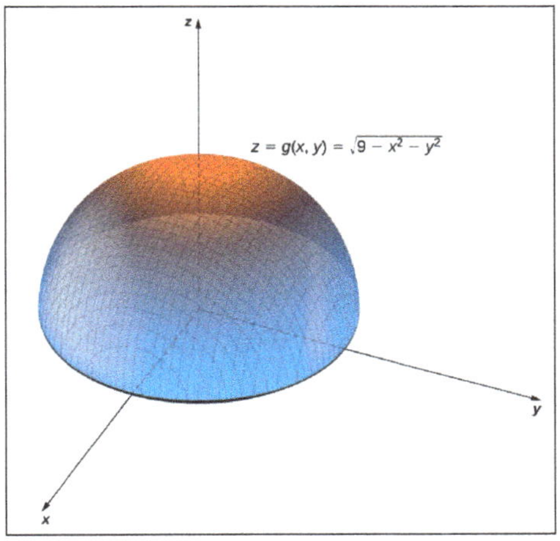

Graph of the hemisphere represented by the given function of two variables.

This function also contains the expression $x^2 + y^2$. Setting this expression equal to various values starting at zero, we obtain circles of increasing radius. The minimum value of $f(x,y) = x^2 + y^2$ is zero (attained when x = y = 0. When x = 0, the function becomes $z = y^2$, and when y = 0, then the function becomes $z = x^2$. These are cross-sections of the graph, and are parabolas. Recall from Introduction to Vectors in Space that the name of the graph of $f(x,y) = x^2 + y^2$ is a paraboloid. The graph of f appears in the following graph:

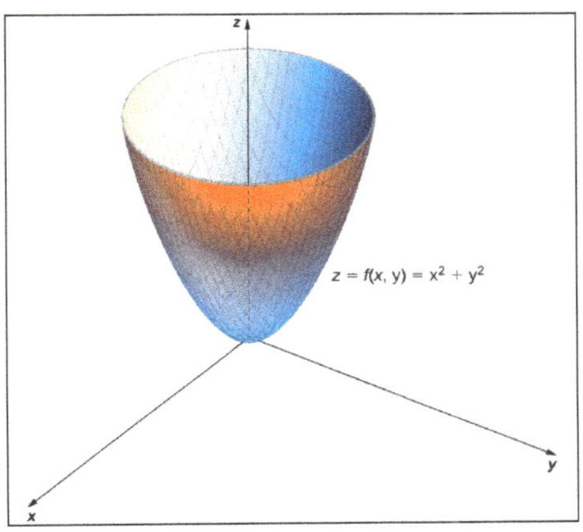

A paraboloid is the graph of the given function of two variables.

Example: Nuts and Bolts. A profit function for a hardware manufacturer is given by,

$$f(x,y)=16-(x-3)^2-(y-2)^2,$$

where x is the number of nuts sold per month (measured in thousands) and y represents the number of bolts sold per month (measured in thousands). Profit is measured in thousands of dollars.

Solution: This function is a polynomial function in two variables. The domain of f consists of (x,y) coordinate pairs that yield a nonnegative profit:

$$16-(x-3)^2-(y-2)^2 \geq 0$$
$$(x-3)^2+(y-2)^2 \leq 16.$$

This is a disk of radius 4 centered at (3,2). A further restriction is that both x and y must be non-negative. When x = 3 and y = 2, f(x,y) = 16. Note that it is possible for either value to be a non-integer; for example, it is possible to sell 2.5 thousand nuts in a month. The domain, therefore, contains thousands of points, so we can consider all points within the disk. For any z < 16, we can solve the equation f(x,y) = 16:

$$16-(x-3)^2-(y-2)^2 = z$$
$$(x-3)^2+(y-2)^2 = 16-z.$$

Since z < 16, we know that $16-z>0$, so the previous equation describes a circle with radius $\sqrt{16-z}$ centered at the point (3,2). Therefore, the range of f(x,y) is $\{z \in \mathbb{R} \mid z \leq 16\}$. The graph of f(x,y) is also a paraboloid, and this paraboloid points downward as shown.

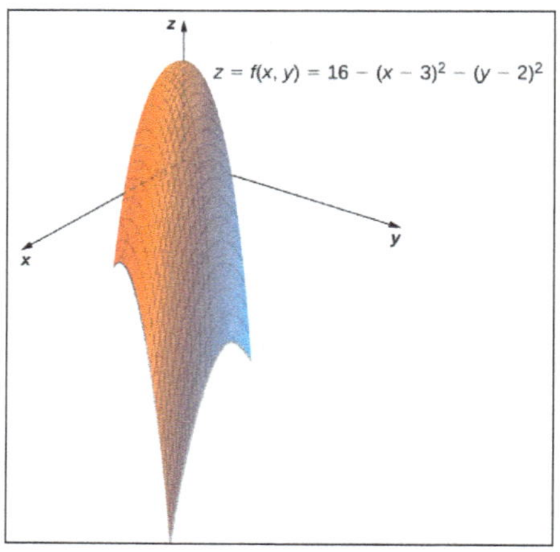

The graph of the given function of two variables is also a paraboloid.

Level Curves

If hikers walk along rugged trails, they might use a topographical map that shows how steeply the trails change. A topographical map contains curved lines called contour lines. Each contour line corresponds to the points on the map that have equal elevation. A level curve of a function of two variables f(x,y) is completely analogous to a contour line on a topographical map.

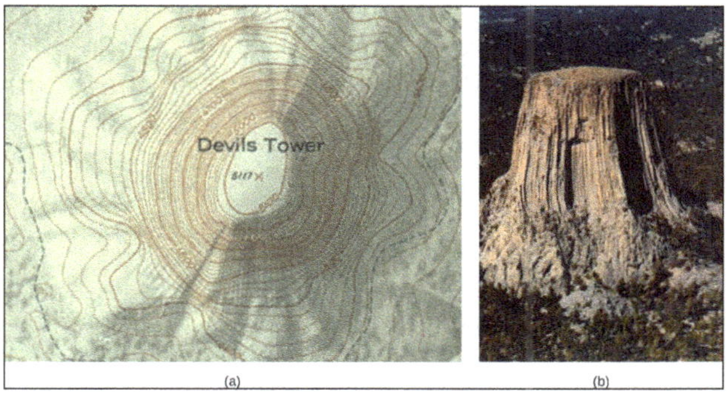

(a) A topographical map of Devil's Tower, Wyoming. Lines that are close together indicate very steep terrain. (b) A perspective photo of Devil's Tower shows just how steep its sides are. The top of the tower has the same shape as the center of the topographical map.

Given a function f(x,y) and a number c in the range of f, a level curve of a function of two variables for the value c is defined to be the set of points satisfying the equation f(x,y) = c.

Returning to the function $g(x,y)=\sqrt{9-x^2-y^2}$ we can determine the level curves of this function. The range of g is the closed interval [0,3]. First, we choose any number in this closed interval—say, c = 2. The level curve corresponding to c = 2 is described by the equation,

$$\sqrt{9-x^2-y^2}=2.$$

To simplify, square both sides of this equation:

$$9-x^2-y^2=4.$$

Now, multiply both sides of the equation by −1 and add 9 to each side:

$$x^2+y^2=5.$$

This equation describes a circle centered at the origin with radius $\sqrt{5}$. Using values of c between 0 and 3 yields other circles also centered at the origin. If c = 3, then the circle has radius 0, so it consists solely of the origin. Figure below is a graph of the level curves of this function corresponding to c = 0,1,2, and 3.

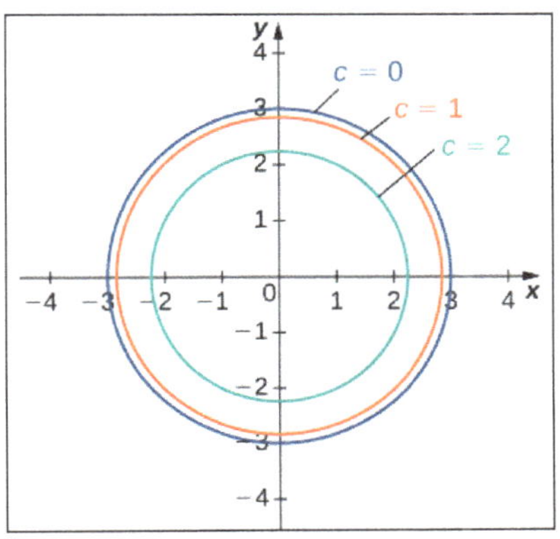

Level curves of the function $g(x,y)=\sqrt{9-x^2-y^2}$ using c = 0,1,2, and 3(c = 3 corresponds to the origin).

A graph of the various level curves of a function is called a contour map.

Example: Making a Contour Map. Given the function $f(x,y)=\sqrt{8+8x-4y-4x^2-y^2}$, find the level curve corresponding to c = 0. Then create a contour map for this function. What are the domain and range of f?

Solution: To find the level curve for c = 0, we set f(x,y) = 0 and solve. This gives,

$$0=\sqrt{8+8x-4y-4x^2-y^2}.$$

We then square both sides and multiply both sides of the equation by −1:

$$4x^2+y^2-8x+4y-8=0.$$

Now, we rearrange the terms, putting the x terms together and the y terms together, and add 8 to each side:

$$4x^2-8x+y^2+4y=8.$$

Next, we group the pairs of terms containing the same variable in parentheses, and factor 4 from the first pair:

$$4(x^2-2x)+(y^2+4y)=8.$$

Then we complete the square in each pair of parentheses and add the correct value to the right-hand side:

$$4(x^2-2x+1)+(y^2+4y+4)=8+4(1)+4.$$

Next, we factor the left-hand side and simplify the right-hand side:

$$4(x-1)^2+(y+2)^2=16.$$

Last, we divide both sides by 16:

$$\frac{(x-1)^2}{4}+\frac{(y+2)^2}{16}=1.$$

This equation describes an ellipse centered at (1,−2). The graph of this ellipse appears in the following graph:

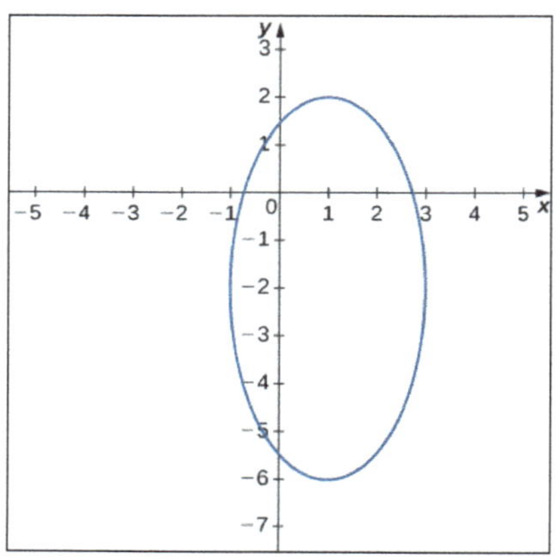

Level curve of the function $f(x,y)=\sqrt{8+8x-4y-4x^2-y^2}$ corresponding to c = 0.

We can repeat the same derivation for values of c less than 4. Then, above equation becomes,

$$\frac{4(x-1)^2}{16-c^2}+\frac{(y+2)^2}{16-c^2}=1$$

for an arbitrary value of c.

Setting the function equal to c, we get:

$$\sqrt{8+8x-4y-4x^2-y^2}=c$$

We then square both sides and multiply both sides of the equation by -1:

$$4x^2+y^2-8x+4y-8=-c^2.$$

Now, we rearrange the terms, putting the x terms together and the y terms together, and add 8 to each side:

$$4x^2-8x+y^2+4y=8-c^2.$$

Next, we group the pairs of terms containing the same variable in parentheses, and factor 4 from the first pair:

$$4(x^2-2x)+(y^2+4y)=8-c^2.$$

Then we complete the square in each pair of parentheses and add the correct value to the right-hand side:

$$4(x^2-2x+1)+(y^2+4y+4)=8-c^2+4(1)+4.$$

Next, we factor the left-hand side and simplify the right-hand side:

$$4(x-1)^2+(y+2)^2=16-c^2.$$

Last, we divide both sides by $16-c^2$:

$$\frac{4(x-1)^2}{16-c^2}+\frac{(y+2)^2}{16-c^2}=1$$

Figure below shows a contour map for f(x,y) using the values c = 0,1,2, and 3. When c = 4, the level curve is the point (−1,2).

Another useful tool for understanding the graph of a function of two variables is called a vertical trace. Level curves are always graphed in the xy-plane, but as their name implies, vertical traces are graphed in the xz− or yz−planes.

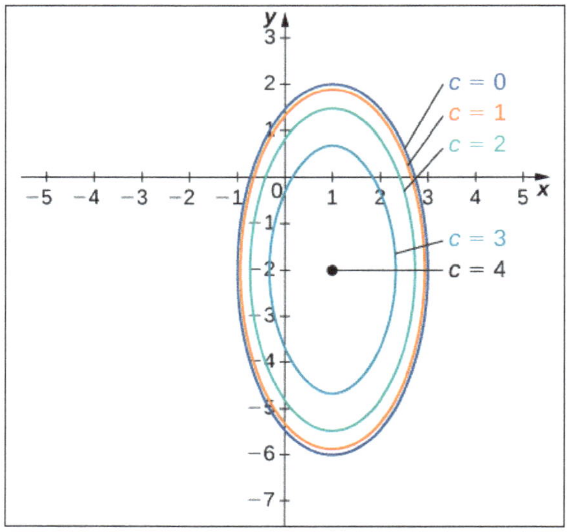

Contour map for the function $f(x,y)=\sqrt{8+8x-4y-4x^2-y^2}$ using the values c = 0,1,2,3, and 4.

Vertical Traces

Consider a function z = f(x,y) with domain D $\subseteq$ IR². A vertical trace of the function can be either the set of points that solves the equation f(a,y) = z for a given constant x=a or f(x,b) = z for a given constant y = b.

Example: Finding Vertical Traces. Find vertical traces for the function f(x,y) = sin x cos y corresponding to $x=-\frac{\pi}{4}, 0, \frac{\pi}{4}$, $y=-\frac{\pi}{4}, 0,$ and $\frac{\pi}{4}$.

Solution: First set $x=-\frac{\pi}{4}$ in the equation z = sin x cos y:

$$z=\sin\left(-\frac{\pi}{4}\right)\cos y=-\frac{\sqrt{2}\cos y}{2}\approx -0.7071\cos y.$$

This describes a cosine graph in the plane $x=-\frac{\pi}{4}$. The other values of z appear in the following table:

Vertical Traces Parallel to the xz–Plane for the Function f(x,y)=sin x cos y.	
c	Vertical Trace for x=cx=c
$-\frac{\pi}{4}$	$z=-\frac{\sqrt{2}\cos y}{2}$
0	z = 0
$\frac{\pi}{4}$	$z=\frac{\sqrt{2}\cos y}{2}$

In a similar fashion, we can substitute the y -values in the equation f(x,y) to obtain the traces in the yz-plane, as listed in the following table:

Vertical Traces Parallel to the yz–Plane for the Function f(x,y) = sin x cos y.	
d	Vertical Trace for y = d
$\frac{\pi}{4}$	$z = \frac{\sqrt{2}\sin x}{2}$
0	z = sin x
$-\frac{\pi}{4}$	$z = \frac{\sqrt{2}\sin x}{2}$

The three traces in the xz-plane are cosine functions; the three traces in the yz-plane are sine functions. These curves appear in the intersections of the surface with the planes $x = -\frac{\pi}{4}$, x = 0, $x = \frac{\pi}{4}$ and $y = -\frac{\pi}{4}$, y = 0, $y = \frac{\pi}{4}$ as shown in the figure.

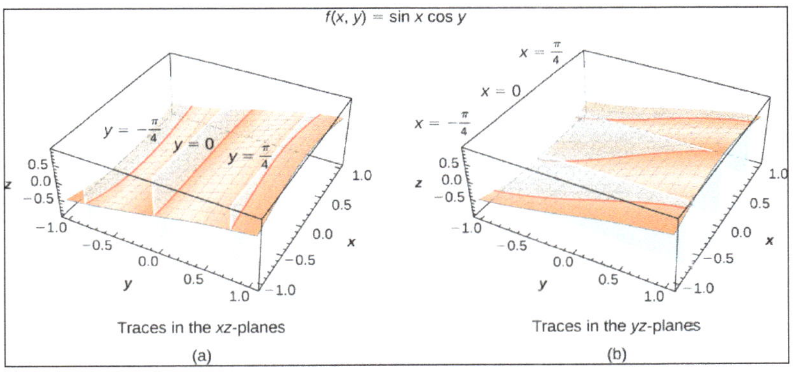

Vertical traces of the function f(x,y) are (a) Cosine curves in the xz-planes and (b) Sine curves in the yz-planes.

Functions of two variables can produce some striking-looking surfaces. Figure below shows two examples.

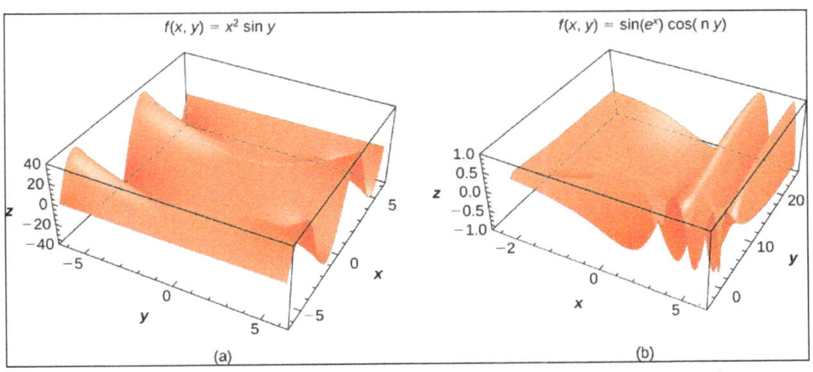

Examples of surfaces representing functions of two variables: (a) A combination of a power function and a sine function and (b) a combination of trigonometric, exponential, and logarithmic functions.

Functions of More than Two Variables

So far, we have examined only functions of two variables. However, it is useful to take a brief look at functions of more than two variables. Two such examples are:

$$\underbrace{f(x,y,z) = x^2 - 2xy + y^2 + 3yz - z^2 + 4x - 2y + 3x - 6}_{\text{a polynomial in three variables}}$$

and

$$g(x,y,t) = (x^2 - 4xy + y^2)\sin t - (3x + 5y)\cos t.$$

In the first function, (x,y,z) represents a point in space, and the function f maps each point in space to a fourth quantity, such as temperature or wind speed. In the second function, (x,y) can represent a point in the plane, and t can represent time. The function might map a point in the plane to a third quantity (for example, pressure) at a given time t. The method for finding the domain of a function of more than two variables is analogous to the method for functions of one or two variables.

Example: Domains for functions of three variables. Find the domain of each of the following functions:

$$f(x,y,z) = \frac{3x - 4y + 2z}{\sqrt{9 - x^2 - y^2 - z^2}}$$

$$g(x,y,t) = \frac{\sqrt{2t - 4}}{x^2 - y^2}$$

Solution: For the function $f(x,y,z) = \frac{3x - 4y + 2z}{\sqrt{9 - x^2 - y^2 - z^2}}$ to be defined (and be a real value), two conditions must hold:

- The denominator cannot be zero.
- The radicand cannot be negative.

Combining these conditions leads to the inequality,

$$9 - x^2 - y^2 - z^2 > 0.$$

Moving the variables to the other side and reversing the inequality gives the domain as,

$$\text{domain}(f) = \{(x,y,z) \in R^3 \mid x^2 + y^2 + z^2 < 9\},$$

which describes a ball of radius 3 centered at the origin. (The surface of the ball is not included in this domain).

For the function $g(x,y,t) = \frac{\sqrt{2t - 4}}{x^2 - y^2}$ to be defined (and be a real value), two conditions must hold:

- The radicand cannot be negative.

- The denominator cannot be zero.

Since the radicand cannot be negative, this implies $2t - 4 \geq 0$, and therefore that $t \geq 2$. Since the denominator cannot be zero, $x^2 - y^2 \neq 0$, or $x^2 \neq y^2$, which can be rewritten as $y = \pm x$, which are the equations of two lines passing through the origin. Therefore, the domain of g is,

$$\text{domain}(g) = \{(x, y, t) \mid y \neq \pm x, t \geq 2\}.$$

Functions of two variables have level curves, which are shown as curves in the xy–plane. However, when the function has three variables, the curves become surfaces, so we can define level surfaces for functions of three variables.

Definition: Level surface of a function of three variables. Given a function f(x,y,z) and a number c in the range of f, a level surface of a function of three variables is defined to be the set of points satisfying the equation f(x,y,z) = c.

Example: Finding a Level Surface. Find the level surface for the function $f(x,y,z) = 4x^2 + 9y^2 - z^2$ corresponding to c = 1.

Solution: The level surface is defined by the equation $4x^2 + 9y^2 - z^2 = 1$. This equation describes a hyperboloid of one sheet as shown in figure.

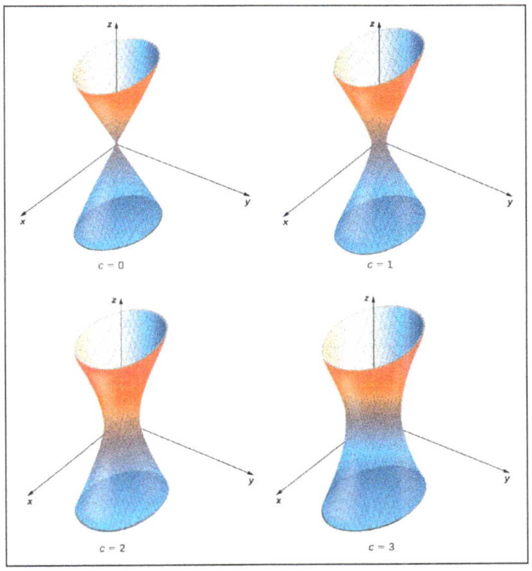

A hyperboloid of one sheet with some of its level surfaces.

Concept of Limits

We know division by zero is not possible in mathematics. If we consider the function definition as,

$$f(x) = \frac{(x^2 - 1)}{(x - 1)}$$

The value of f(x) at x=1 is indeterminate.

$$f(1) = \frac{(1^2 - 1)}{(1 - 1)} = \frac{0}{0}$$

More simply, the value of the function f(x) does not exist at x=1. So, instead of x=1 we consider values of x sufficiently close to 1, i.e., as close to 1 as possible.

X	F(X)
0.5	1.50000
0.9	1.90000
0.99	1.99000
0.999	1.99900
0.9999	1.99990
0.99999	1.99999
...	...

We can see that as the values of x approach 1, the value of the function f(x) approach 2. In the table we have stopped at 0.99999, but if we take values of x even closer to 1, the corresponding values of f(x) will be even closer to 2. Here, as the values of x increase towards 1 the values of f(x) increase towards 2. This is symbolically written as:

$$\lim_{x \to 1+} f(x) = 2$$

It reads, "limit x tends to 1 plus f(x) is equal to 2". It is to be noted that the '+' sign signifies values of x greater than 1 and not "positive" values of x. Let us now look at the following table:

X	F(X)
1.5	2.50000
1.1	2.10000
1.01	2.01000
1.001	2.00100
1.0001	2.00010
1.00001	2.00001
...	...

We can see that as the values of x come closer and closer to 1, the corresponding values of f(x) come closer and closer to 2. In other words, as the values of x approach 1, the corresponding values of f(x) approach 2. The only difference is that here the values of x decrease towards 1 and the values of f(x) decrease towards 2. This is symbolically written as:

$$\lim_{x \to 1-} f(x) = 2$$

It reads, "limit x tends to 1 minus f(x) is equal to 2". It is to be noted that the '-' sign signifies values of x less than 1 and not "negative" values of x.

In practice, the numerical difference between the value x=1 and a value of x sufficiently close to 1 (such as, x=1.0000001 or x=0.9999999) can be made as small as we please and hence can be neglected. Similarly, the numerical difference between the value f(x)=2 and a value of f(x) very close to 2 can be made as small as we please and hence be neglected if the value of x is taken sufficiently close to 1.

In general, when the value of f(x) cannot be determined for a particular value of x, say, x=a then there may exist a definite finite number b, such that the value of f(x) gradually tends to that finite number b when x tends to a. However we cannot say whether that finite number b will always exist or not. From this observation, the concept of limit has been developed by mathematicians.

Limit of a Variable

Let us consider a real variable x. Let a be a constant. Then by 'x tends to a' we mean x successively assumes values either greater than or less than a and the numerical difference between the assumed value of x, i.e., |x - a| becomes smaller and smaller. In this case, x becomes very close to a (but $x \neq a$) and we say 'x approaches a'.

- If x approaches an assuming values greater than a then we say 'x tends to a from the right side' and we denote it by $x \to a^+$.
- If x approaches an assuming values less than a then we say 'x tends to a from the left side' and we denote it by $x \to a^-$.

Limiting Value of a Function

We assume x to be a real variable, a is a real constant and f(x) is a single-valued function of x. If x gradually approaches a assuming values which are greater than a and if the corresponding values of f(x) exist and these values gradually approach a finite constant l_1, then l_1, is called the right hand limiting value of f(x) or the right hand limit of f(x) and it is denoted by,

$$\lim_{x \to a+} f(x) = l_1$$

Again, if x gradually approaches a assuming values which are less than a and if the corresponding values of f(x) exist and these values gradually approach a finite constant l_2 then l_2 is called the left hand limiting value of f(x) or the left hand limit of f(x) and it is denoted by,

$$\lim_{x \to a-} f(x) = l_2$$

When x approaches a assuming values either greater than or less than a and f(x) assumes finite values for every value of x and if those values of f(x) gradually approach a finite constant l, then l is called the limiting value of f(x). It is denoted by,

$$\lim_{x \to a} f(x) = l$$

The limit of the function $l = \lim_{x \to a} f(x)$ f(x) exists only if both l_1 (the right hand limit) and l_2 (the left hand limit) exist and $l_1 = l_2$, i.e., if,

$$\lim_{x \to a+} f(x) = \lim_{x \to a-} f(x)$$

l does not exist if,

- l_1 is indeterminate OR.
- l_2 is indeterminate OR.
- $l_1 \neq l_2$.

Different Types of Limits

The Limit of a Function as x Tends to Infinity

If we have a sequence $(y_n)_{n=1}^{\infty}$, we can say what it means for the sequence to have a limit as n tends to infinity. We write,

$$y_n \to l \text{ as } n \to \infty$$

if, however small a distance we choose, y_n eventually gets closer to l than that distance, and stays closer. We can also write,

$$\lim_{x \to \infty} f(x) = l.$$

For example, consider the sequence where $y_n = 1/n$. The numbers in this sequence get closer and closer to zero. Whatever positive number we choose, y_n will eventually become smaller than that number, and stay smaller. So y_n eventually gets closer to zero than any distance we choose, and stays closer. We say that the sequence has limit zero as n tends to infinity.

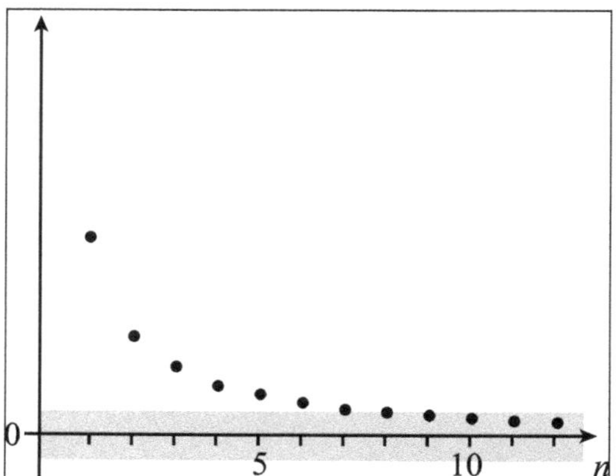

the sequence $\left(\frac{1}{n}\right)_{n=1}^{\infty}$

We define the limit of a function in a similar way. For example, the points of the sequence $(1/n)_{n=1}^{\infty}$ are also points on the graph of the function f(x) = 1/x for x > 0. As x gets larger, f(x) gets closer and closer to zero. In fact, f(x) will get closer to zero than any distance we choose, and will stay closer. We say that f(x) has limit zero as x tends to infinity, and we write,

$$f(x) \to 0 \text{ as } x \to \infty, \text{ or } \lim_{x \to \infty} f(x) = 0.$$

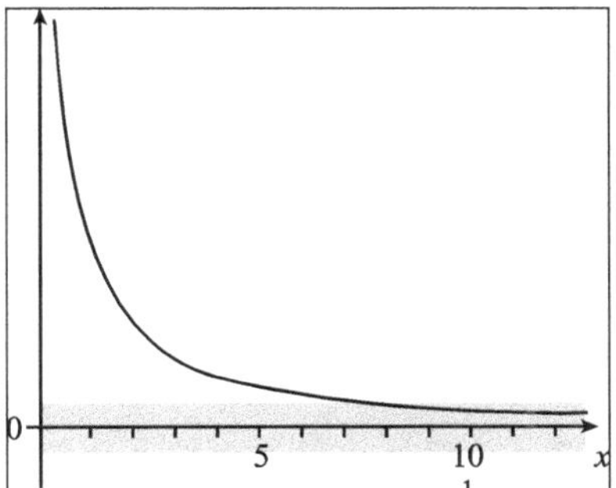

the function $f(x) = \frac{1}{x}$

Another example of a function that has a limit as x tends to infinity is the function $f(x) = 3-1/x^2$ for $x > 0$. As x gets larger, f(x) gets closer and closer to 3. For any small distance, f(x) eventually gets closer to 3 than that distance, and stays closer. So we say that f(x) has limit 3 as x tends to infinity, and we write,

$$f(x) \to 3 \text{ as } x \to \infty, \text{ or } \lim_{x\to\infty} f(x) = 3.$$

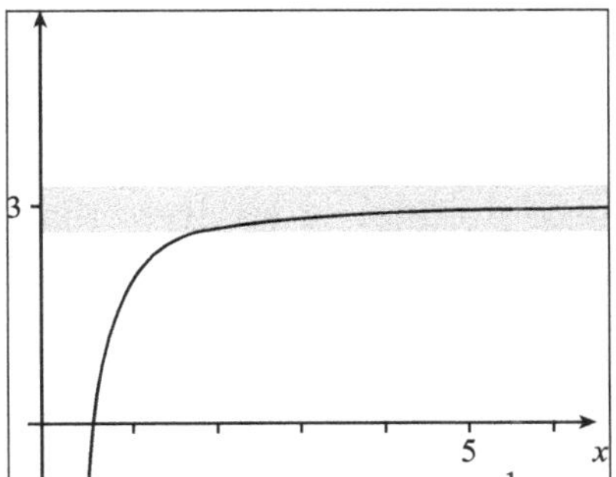

the function $f(x) = 3 - \frac{1}{x^2}$

In general, we say that f(x) tends to a real limit l as x tends to infinity if, however small a distance we choose, f(x) gets closer than that distance to l and stays closer as x increases. Of course, not all functions have real limits as x tends to infinity. Let us look at some other types of behaviour. If we take the function $f(x) = x^2$, we see that f(x) does not get closer to any particular number as x increases. Instead, f(x) just gets larger and larger. At some point, f(x) will get larger than any number we choose, and will stay larger. In this case, we say that f(x) tends to infinity as x tends to infinity, and we write,

$$f(x) \to \infty \text{ as } x \to \infty, \text{ or } \lim_{x\to\infty} f(x) = \infty.$$

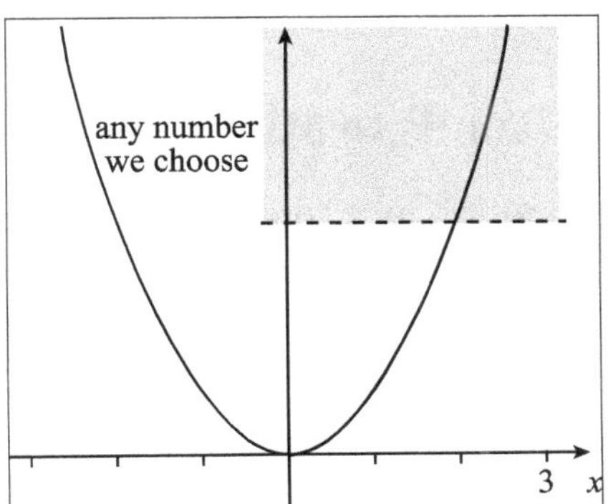

the function $f(x) = x^2$

The function f(x) = −x does not have a real limit as x tends to infinity. As x gets larger, this function eventually gets more negative than any number we can choose, and it will stay more negative. In this case, we say that f(x) tends to minus infinity as x tends to infinity, and we write,

$$f(x) \to -\infty \text{ as } x \to \infty, \text{ or } \lim_{x\to\infty} f(x) = -\infty.$$

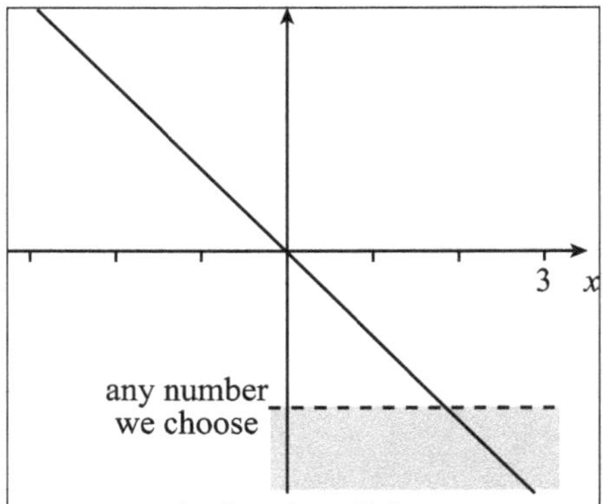

the function $f(x) = -x$

Some functions do not have any kind of limit as x tends to infinity. For example, consider the function f(x) = x sin x. This function does not get close to any particular real number as x gets large, because we can always choose a value of x to make f(x) larger than any number we choose. However f(x) does not tend to infinity, because it does not stay larger than the number we have chosen, but instead returns to zero. For a similar reason, f(x) does not tend to minus infinity. So we cannot talk about the limit of this function as x tends to infinity.

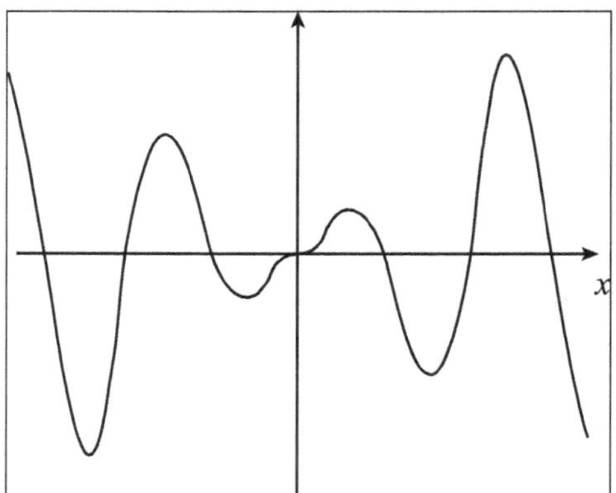

the function $f(x) = x \sin x$

The Limit of a Function as x Tends to Minus Infinity

As well as defining the limit of a function as x tends to infinity, we can also define the limit as x tends to minus infinity. Consider the function f(x) = e^x. As x becomes more and more negative, f(x) gets closer and closer to zero. However small a distance we choose, f(x) gets closer than that distance to zero, and it stays closer as x becomes more negative. We say that f(x) has limit zero as x tends to minus infinity, and we write,

$$f(x) \to 0 \text{ as } x \to -\infty, \text{ or } \lim_{x\to-\infty} f(x) = 0.$$

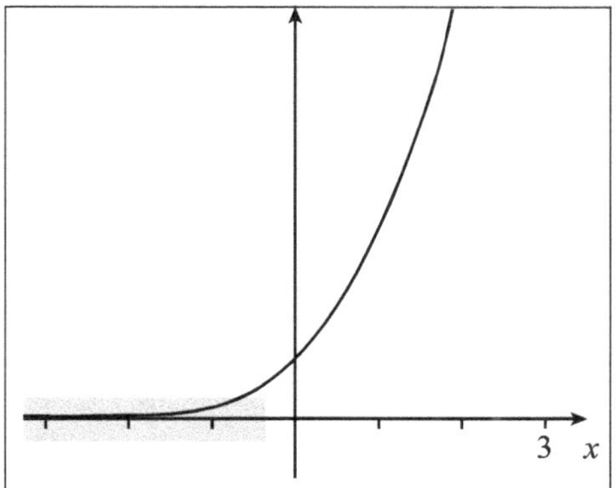

the function $f(x) = e^x$

In general we write,

$$f(x) \to l \text{ as } x \to -\infty, \text{ or } \lim_{x\to-\infty} f(x) = l.$$

If, however small a distance we choose, f(x) eventually gets closer to l than that distance, and stays closer, as x becomes large and negative.

If, as x gets more negative, a function gets larger and stays larger than any number we can choose, we write,

$$f(x) \to \infty \text{ as } x \to -\infty, \text{ or } \lim_{x\to-\infty} f(x) = \infty.$$

For example, take the function $f(x) = x^2$ again. It tends to infinity as x tends to infinity. But it also tends to infinity as x tends to minus infinity. As x gets large and negative, the function gets larger than any number we can choose, and stays larger.

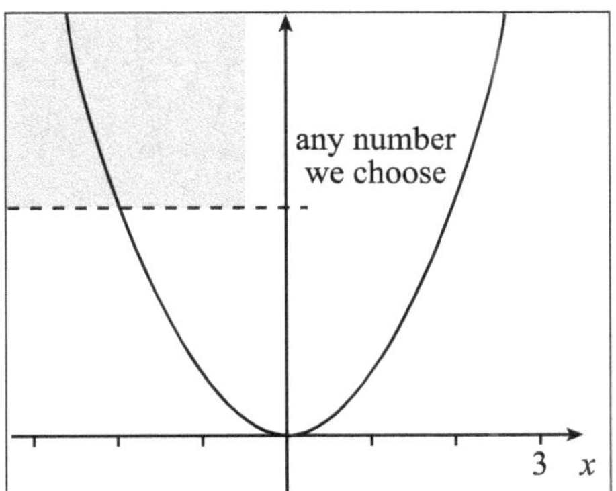

the function $f(x) = x^2$

If, instead, as x gets more negative, a function gets more negative and stays more negative than any number we can choose, we write,

$$f(x) \to -\infty \text{ as } x \to -\infty, \text{ or } \lim_{x\to-\infty} f(x) = -\infty.$$

As an example, consider the function $f(x) = x^3$. As x gets more and more negative, x^3 becomes more negative than any number we can choose, and stays more negative. So f(x) tends to minus infinity as x tends to minus infinity.

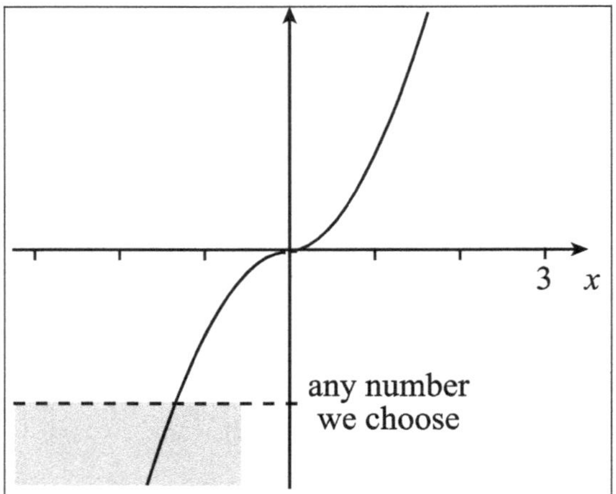

the function $f(x) = x^3$

Some functions do not have any kind of limit as x tends to minus infinity. For example, consider the function f(x) = x sin x. This function does not get close to any particular real number as x gets large and negative, because we can always choose a value of x to make f(x) larger than any number we choose. However f(x) does not tend to infinity, because it does not stay larger than the number we have chosen as we decrease x, but instead returns to zero. For a similar reason, f(x) does not tend to minus infinity. So we cannot talk about the limit of this function as x tends to minus infinity.

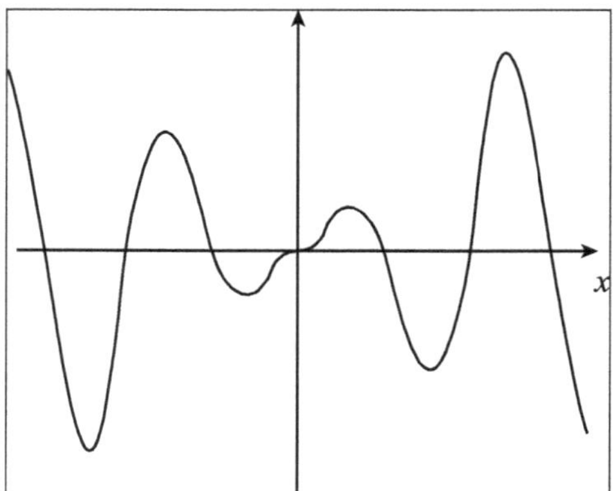

the function $f(x) = x \sin x$

The Limit of a Function as x Tends to a Real Number

There is one more type of limit that we can define for functions. Let us consider the function f(x) = x+3. If we choose a number, such as 1, then as x gets closer and closer to that number, f(x) also gets closer and closer to a number, in this case 4. We write,

$$f(x) \to 4 \text{ as } x \to 1, \text{ or } \lim_{x \to 1} f(x) = 4.$$

Similarly, f(x) gets closer and closer to 8 as x gets closer and closer to 5. So we write,

$$f(x) \to 8 \text{ as } x \to 5, \text{ or } \lim_{x \to 5} f(x) = 8.$$

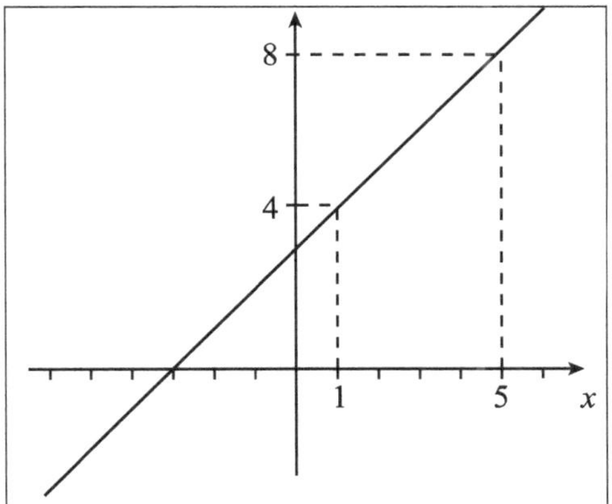

the function $f(x) = x + 3$

Now this definition of a limit might not look very useful. We know that when x = 1 then the value of f(x) is 4. And again, when x = 5 then the value of (x) is 8. Why would we want to bother looking at what happens when x gets closer and closer to these numbers? The reason is that we might sometimes have a function that is not defined at a point. For example, consider the graph of the function x sin(1/x). This function is defined for every value apart from zero, because at x = 0 we have a fraction with a zero denominator inside the sine function. But if we look at the rest of the graph, we can see that f(x) gets closer and closer to zero as x gets closer and closer to zero. So we write,

$$f(x) \to 0 \text{ as } x \to 0, \text{ or } \lim_{x \to 0} f(x) = 0.$$

We might say that f(0) 'ought to equal 0', even though there is no properly defined value for f(x) at zero.

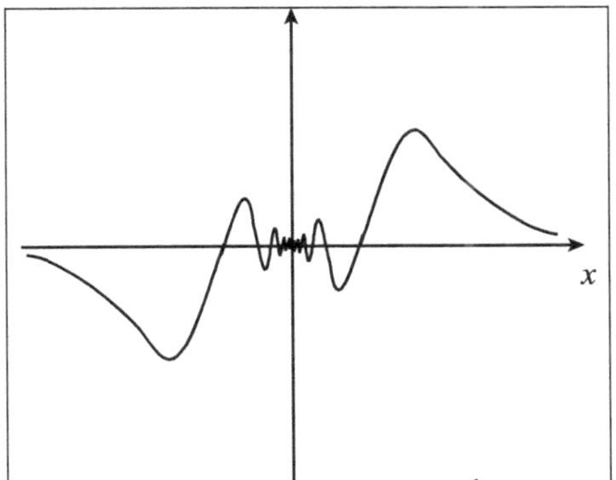

the function $f(x) = x \sin \frac{1}{x}$

So the limit of a f(x) at a point where the function is undefined can be thought of as the value the function should be taking at that point.

A similar thing happens with the function $f(x) = e^{-1/x^2}$. This function gets closer and closer to zero as x gets closer and closer to zero. But again the value of the function at x = 0 is not defined, because there is a fraction with a zero denominator in the function. Nevertheless, we can see that the limit of the function is 0 as x tends to zero, so we write

$$f(x) \to 0 \text{ as } x \to 0, \text{ or } \lim_{x \to 0} f(x) = 0.$$

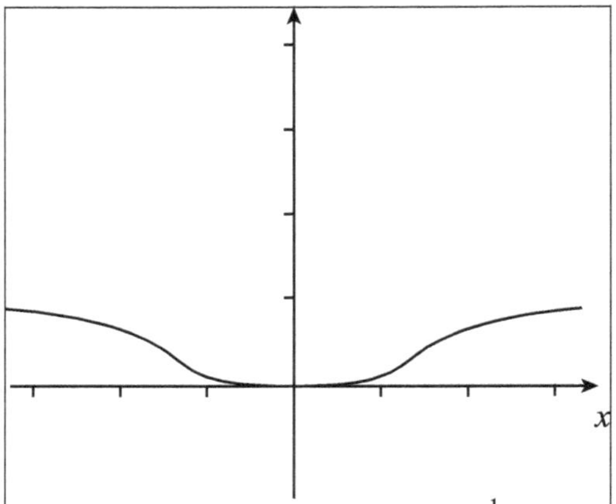

the function $f(x) = e^{\frac{1}{x^2}}$

Some functions do not have limits at certain points. If we take the function f(x) = |x|/x then, for x > 0, f(x) = x/x = 1. But for x < 0, f(x) = −x/x = −1. At x = 0 the function is undefined, because there is a zero denominator. If x is positive then going closer and closer to zero keeps f(x) at 1. But if x is negative, going closer and closer to zero keeps f(x) at −1. So this function does not have a limit at x = 0.

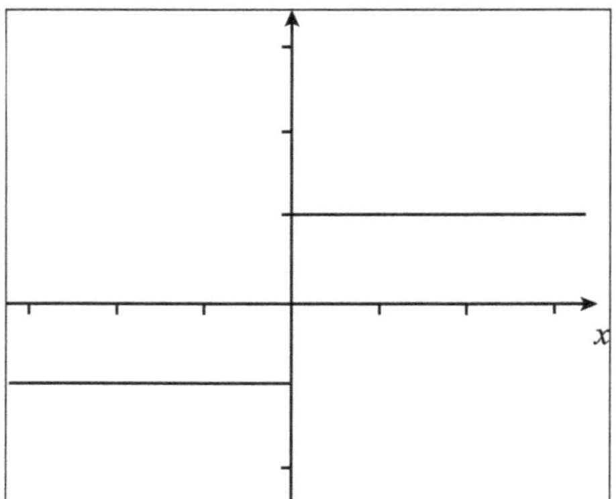

the function $f(x) = \frac{|x|}{x}$

Some More Limits

Limits at Large Numbers

	Definition in Words	Precise Mathematical Definition
Large POSITIVE numbers	Let f be a function defined on some interval (a, ∞). Then $\lim_{x \to \infty} f(x) = L$ means that the values of f(x) can be made arbitrarily close to L by taking x sufficiently large in a positive direction.	Let f be a function defined on some interval (a, ∞). Then $\lim_{x \to \infty} f(x) = L$ if for every ε > 0 there is a corresponding number N such that if x > N then |f(x) − L|< ε.
Large NEGATIVE numbers	Let f be a function defined on some interval (-∞,a).∞). Then $\lim_{x \to -\infty} f(x) = L$ means that the values of f(x) can be made arbitrarily close to L by taking x sufficiently large in a negative direction.	Let f be a function defined on some interval (-∞,a). Then $\lim_{x \to -\infty} f(x) = L$ if for every ε > 0 there is a corresponding number N such that if x < N then |f(x) − L|< ε.

Horizontal Asymptote

The line y = L is a horizontal asymptote of the curve y = f(x) if either is true:

- $\lim_{x \to \infty} f(x) = L$ or
- $\lim_{x \to -\infty} f(x) = L$

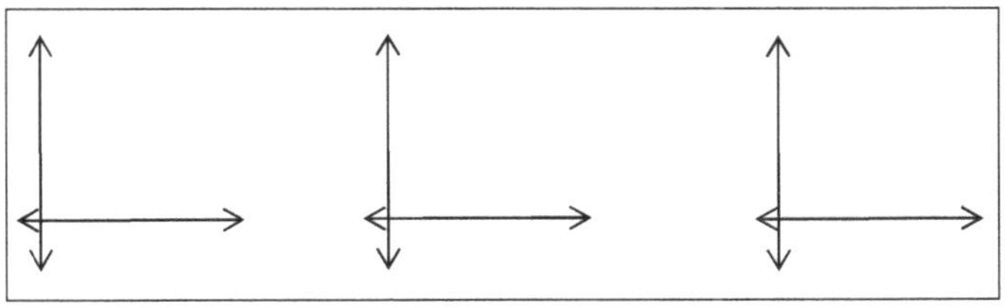

Vertical Asymptote

The line x = a is a vertical asymptote of the curve y = f(x) if at least one of the following is true:

- $\lim_{x \to a} f(x) = \infty$
- $\lim_{x \to a^-} f(x) = \infty$
- $\lim_{x \to a^+} f(x) = \infty$
- $\lim_{x \to a} f(x) = -\infty$
- $\lim_{x \to a^-} f(x) = -\infty$
- $\lim_{x \to a^+} f(x) = -\infty$

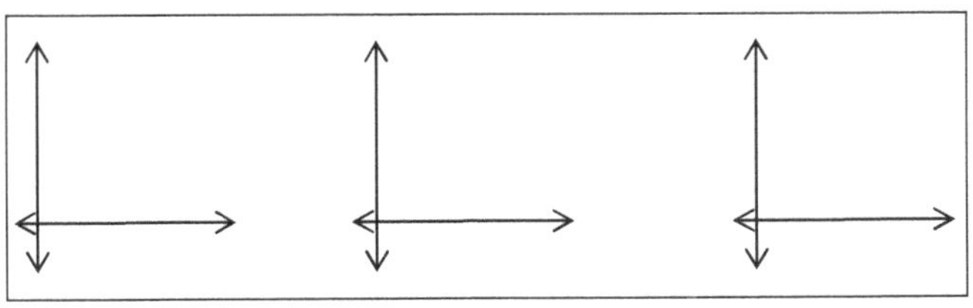

Theorem:

- If r > 0 is a rational number then $\lim_{x \to \infty} \frac{1}{x^r} = 0$.
- If r > 0 is a rational number such that x^r is defined for all x then $\lim_{x \to -\infty} \frac{1}{x^r} = 0$.

Indeterminate Forms

1. Indeterminate Forms $\frac{0}{0}$: Let f(x) and g(x) be two functions such that,

$\lim_{x \to a} f(x) = 0$ and $\lim_{x \to a} g(x) = 0$.

Then the function $\frac{f(x)}{g(x)}$ has the indeterminate form $\frac{0}{0}$ at x = a. To find the limit at x = a when the function $\frac{f(x)}{g(x)}$ has the indeterminate form $\frac{0}{0}$ at this point, we must factor the numerator and denominator and then reduce the terms that approach zero.

2. Indeterminate Forms $\frac{\infty}{\infty}$: Let f(x) and g(x) be two functions such that,

$$\lim_{x \to a} f(x) = \pm\infty \text{ and } \lim_{x \to a} g(x) = \pm\infty.$$

where a where a is a real number, or + ∞ or −∞.

It is said that the function $\frac{f(x)}{g(x)}$ has the indeterminate form $\frac{\infty}{\infty}$ at this point. To find the limit, we must divide the numerator and denominator by x of highest degree.

3. Indeterminate Forms $\infty - \infty, 0 \cdot \infty, \infty^0, 1^\infty$: Indeterminate forms of these types can usually be treated by putting them into one of the forms $\frac{0}{0}$ or $\frac{\infty}{\infty}$.

L' Hopital's Rule

Following are two of the forms of l'Hopital's Rule:

Theorem: (l'Hopital's Rule for zero over zero): Suppose that $\lim_{x \to a} f(x) = 0$, $\lim_{x \to a} g(x) = 0$, and that functions f and g are differentiable on an open interval I containing a. Assume also that $g'(x) \neq 0$ in I if $x \neq a$. Then,

$$\lim_{x \to a} \frac{f(x)}{g(x)} = \lim_{x \to a} \frac{f'(x)}{g'(x)}$$

so long as the limit is finite, +∞, or −∞. Similar results hold for $x \to \infty$ and $x \to -\infty$.

Theorem: (l'Hopital's Rule for infinity over infinity): Assume that functions f and g are differentiable for all x larger than some fixed number. If $\lim_{x \to a} f(x) = \infty$ and $\lim_{x \to a} g(x) = \infty$, then,

$$\lim_{x \to a} \frac{f(x)}{g(x)} = \lim_{x \to a} \frac{f'(x)}{g'(x)}$$

so long as the limit is finite, +∞, or −∞. Similar results hold for $x \to \infty$ and $x \to -\infty$.

In both forms of l'Hopital's Rule it should be noted that you are required to differentiate (separately) the numerator and denominator of the ratio if either of the indeterminate forms $\frac{"0"}{0}$ or $\frac{"\infty"}{\infty}$ arises in the computation of a limit. Do not confuse l'Hopital's Rule with the Quotient Rule for derivatives.

Example:

$$\lim_{x\to 2}\frac{x-2}{x^2-4}=\frac{"2-2"}{(2)^2-4}=\frac{"0"}{0}$$

(Now apply Theorem: l'Hopital's Rule for zero over zero. Differentiate top and bottom separately).

$$=\lim_{x\to 2}\frac{1-0}{2x-0}$$

$$=\lim_{x\to 2}\frac{1}{2x}$$

$$=\frac{1}{2(2)}$$

$$=\frac{1}{4}$$

Example:

$$\lim_{x\to\infty}\frac{2x+7}{3x^2-5}=\frac{"\infty"}{\infty}=\frac{"0"}{0}$$

(Now apply Theorem: l'Hopital's Rule for infinity over infinity. Differentiate top and bottom separately).

$$=\lim_{x\to\infty}\frac{2+1}{6x-0}$$

$$=\lim_{x\to\infty}\frac{1}{3x}$$

$$=\frac{"1"}{\infty}$$

$$=0$$

Indeterminate forms besides $\frac{"0"}{0}$ and $\frac{"\infty"}{\infty}$ include "$0\cdot\infty$, "$\infty-\infty$", "1^∞", "0^0", and "∞^0". These forms also arise in the computation of limits and can often be algebraically transformed into the form $\frac{"0"}{0}$ or $\frac{"\infty"}{\infty}$ so that l'Hopital's Rule can be applied. Following are two examples of such transformations. The second example uses the fact that $y = e^x$ and $y = \ln x$ are inverse functions, so that $z = e^{\ln z}$ for all $z > 0$ and $\ln z^m = m \ln z$ for all $z > 0$ and any m.

Example:

$$\lim_{x\to 0^+}\sqrt{x}\cdot\ln x="0\cdot-\infty"$$

(Circumvent this indeterminate form by "flipping" $\sqrt{x}$).

$$=\lim_{x\to 0^+}\frac{\ln x}{1/\sqrt{x}}=\frac{"-\infty"}{\infty}$$

(Now use Theorem: l'Hopital's Rule for infinity over infinity for l'Hopital's Rule).

$$= \lim_{x \to 0^+} \frac{1/x}{-1/2x^{3/2}}$$

$$= \lim_{x \to 0^+} -2\sqrt{x}$$

$$= -2\sqrt{0}$$

$$= -2(0)$$

$$= 0$$

Example:

$$\lim_{x \to 0^+} x^x = "0^0"$$

(Use the fact that $z = e^{\ln z}$).

$$= \lim_{x \to 0^+} e^{\ln x^x}$$

(Use the fact that $\ln z^m = m \ln z$).

$$= \lim_{x \to 0^+} e^{x \ln x}$$

(This next step uses the fact that $y = e^{x \ln x}$ is a continuous function).

$$= e^{\lim_{x \to 0^+} x \ln x}$$

$$= e^{"0 \cdot (-\infty)"}$$

(Circumvent this indeterminate form by "flipping" x).

$$= e^{\lim_{x \to 0^+} \frac{\ln x}{1/x}}$$

(Now apply Theorem: l'Hopital's Rule for infinity over infinity for l'Hopita's Rule).

$$= e^{\lim_{x \to 0^+} \frac{1/x}{-1/x^2}}$$

$$= e^{\lim_{x \to 0^+} (-x)}$$

$$= e^0$$

$$= 1$$

References

- Definition-of-calculus-2311607: thoughtco.com, Retrieved 02, July 2020
- Functions-and-its-types: byjus.com, Retrieved 21, March 2020
- Implicit-and-explicit-functions: sciencehq.com, Retrieved 13, August 2020
- Inverse-functions-intro: mesacc.edu, Retrieved 20, May 2020
- Definition-graphs-trigonometric-functions: math24.net, Retrieved 06, February 2020
- A-Functions-of-Multiple-Variables-and-Partial-Derivatives: libretexts.org, Retrieved 11, June 2020

2

Differential Calculus

The area of study which deals with the definition, properties and applications of the derivative of a function is called differential calculus. The process through which the derivative is found is termed as differentiation. The topics elaborated in this chapter will help in gaining a better perspective about differential calculus as well as the rules of differentiation.

In day to day life we are often interested in the extent to which a change in one quantity affects a change in another related quantity. This is called a rate of change. For example, if you own a motor car you might be interested in how much a change in the amount of fuel used affects how far you have travelled. This rate of change is called fuel consumption. If your car has high fuel consumption then a large change in the amount of fuel in your tank is accompanied by a small change in the distance you have travelled.

Sprinters are interested in how a change in time is related to a change in their position. This rate of change is called velocity. Other rates of change may not have special names like fuel consumption or velocity, but are nonetheless important. For example, an agronomist might be interested in the extent to which a change in the amount of fertiliser used on a particular crop affects the yield of the crop. Economists want to know how a change in the price of a product affects the demand for that product. Differential calculus is about describing in a precise fashion the ways in which related quantities change.

Differentiation from First Principle

Differentiating a Linear Function

A straight line has a constant gradient, or in other words, the rate of change of y with respect to x is a constant.

Example: Consider the straight line $y = 3x + 2$ shown below.

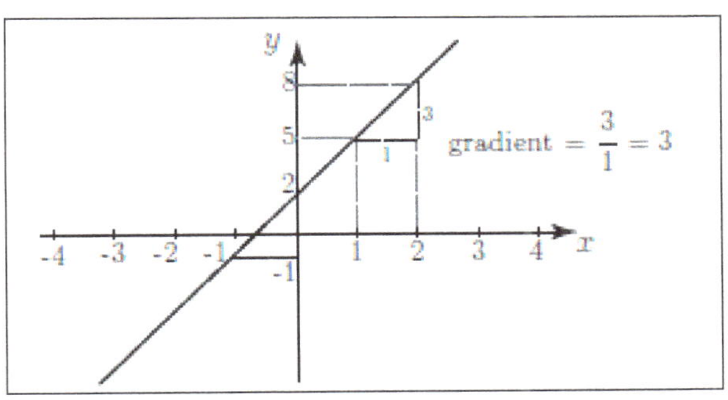

A graph of the straight line y = 3x + 2.

We can calculate the gradient of this line as follows. We take two points and calculate the change in y divided by the change in x. When x changes from −1 to 0, y changes from −1 to 2, and so,

$$\text{the gradient} = \frac{2-(-1)}{0-(-1)} = \frac{3}{1} = 3$$

No matter which pair of points we choose the value of the gradient is always 3. Values of the function y = 3x + 2 are shown below,

x	−3	−2	−1	0	1	2	3
3x	−9	−6	−3	0	3	6	9
2	2	2	2	2	2	2	2
y = 3x + 2	−7	−4	−1	2	5	8	11

Look at the table of values and note that for every unit increase in x we always get an increase of 3 units in y. In other words, y increases as a rate of 3 units, for every unit increase in x. We say that "the rate of change of y with respect to x is 3".

Observe that the gradient of the straight line is the same as the rate of change of y with respect to x. For a straight line: the rate of change of y with respect to x is the same as the gradient of the line.

Differentiation from First Principles of some Simple Curves

For any curve it is clear that if we choose two points and join them, this produces a straight line. For different pairs of points we will get different lines, with very different gradients.

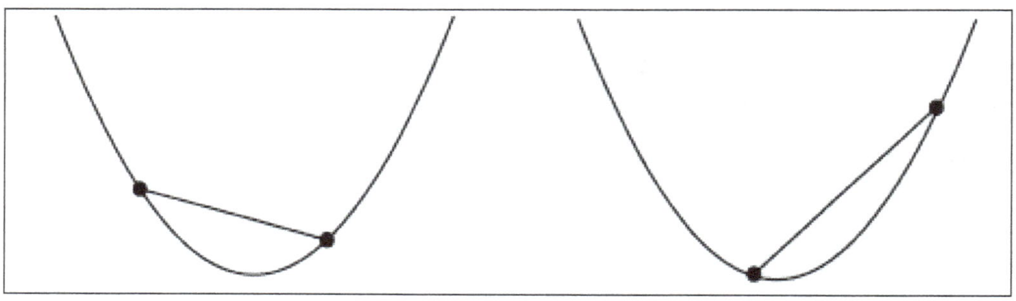

Joining different pairs of points on a curve produces lines with different gradients.

Example: Suppose we look at y = x^2.

x	−3	−2	−1	0	1	2	3
$y = x^2$	9	4	1	0	1	4	9

Note that as x increases by one unit, from −3 to −2, the value of y decreases from 9 to 4. It has reduced by 5 units. But when x increases from −2 to −1, y decreases from 4 to 1. It has reduced by 3. So even for a simple function like y = x^2. y is not changing constantly with x. The rate of change of y with respect to x is not a constant.

Calculating the Rate of Change at a Point

We now explain how to calculate the rate of change at any point on a curve $y = f(x)$. This is defined to be the gradient of the tangent drawn at that point.

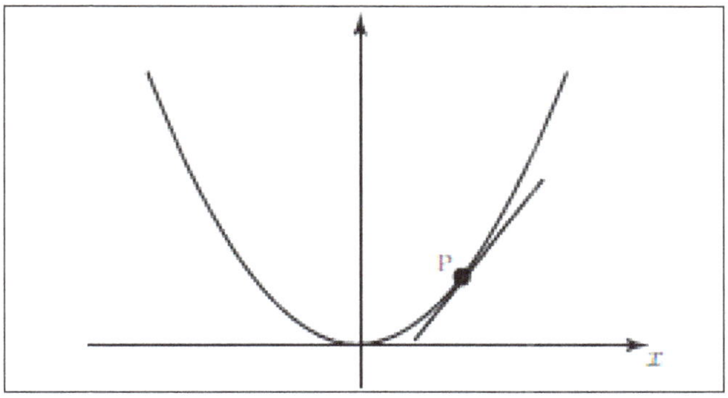

The rate of change at a point P is defined to be the gradient of the tangent at P. Note: The gradient of a curve $y = f(x)$ at a given point is defined to be the gradient of the tangent at that point.

We use this definition to calculate the gradient at any particular point. Consider the graph below which shows a fixed point P on a curve. We also show a sequence of points Q_1, Q_2, . . . getting closer and closer to P. We see that the lines from P to each of the Q's get nearer and nearer to becoming a tangent at P as the Q's get nearer to P.

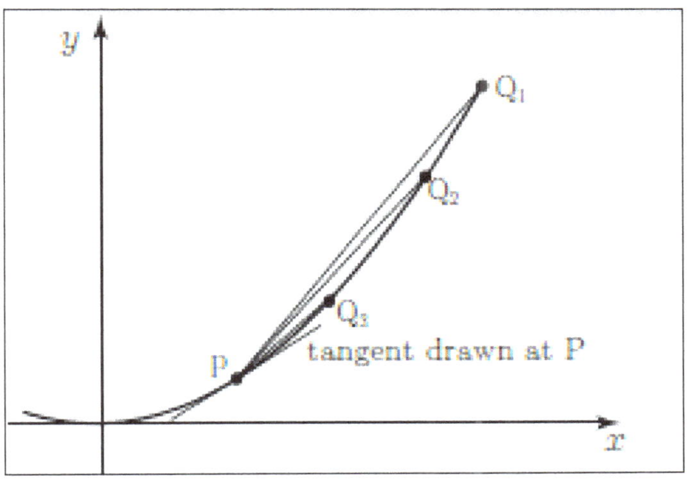

The lines through P and Q approach the tangent at P when Q is very close to P. So if we calculate the gradient of one of these lines, and let the point Q approach the point P along the curve, then the gradient of the line should approach the gradient of the tangent at P, and hence the gradient of the curve.

Example: We shall perform the calculation for the curve $y = x^2$ at the point, P, where $x = 3$. The graph below shows the graph of $y = x^2$ with the point P marked. We choose a nearby point Q and join P and Q with a straight line. We will choose Q so that it is quite close to P. Point R is vertically below Q, at the same height as point P, so that PQR is right-angled.

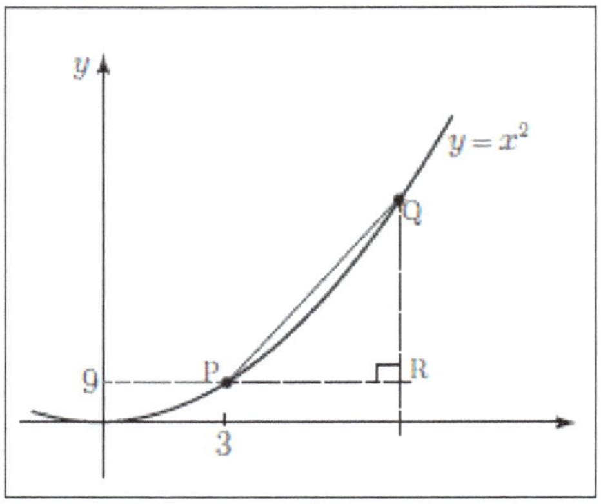

The graph of y = x^2. P is the point (3, 9). Q is a nearby point.

Suppose we choose point Q so that PR = 0.1. The x coordinate of Q is then 3.1 and its y coordinate is 3.1^2. Knowing these values we can calculate the change in y divided by the change in x and hence the gradient of the line PQ.

Gradient of PQ = $\dfrac{QR}{PR} = \dfrac{3.1^2 - 3^2}{0.1} = 6.1$

We can take the gradient of PQ as an approximation to the gradient of the tangent at P, and hence the rate of change of y with respect to x at the point P.

The gradient of PQ will be a better approximation if we take Q closer to P. The table below shows the effect of reducing PR successively, and recalculating the gradient.

PR	0.1	0.01	0.001	0.0001
QR	0.61	0.0601	0.006001	0.00060001
$\dfrac{QR}{PR}$	6.1	6.01	6.001	6.0001

The gradient of the line PQ, QR/PR seems to approach 6 as Q approaches P. Observe that as Q gets closer to P the gradient of PQ seems to be getting nearer and nearer to 6. We will now repeat the calculation for a general point P which has coordinates (x, y).

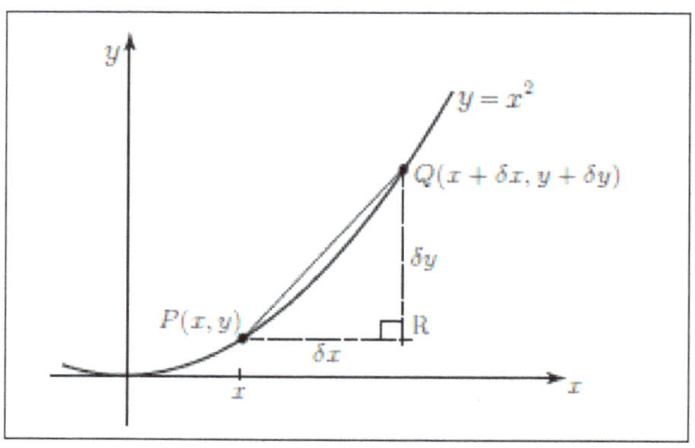

The graph of y = x^2. P is the point (x, y). Q is a nearby point. Point Q is chosen to be close to P on the curve. The x coordinate of Q is x + dx where dx is the symbol we use for a small change, or small increment in x. The corresponding change in y is written as dy. So the coordinates of Q are (x + dx, y + dy). Because we are considering the graph of $y = x^2$, we know that $y + dy = (x + dx)^2$.

$$Y+\delta y=(x+\delta x)^2$$
$$=x^2+2x(\delta x)+(\delta x)^2$$

But $y=x^2$ and so,

$$\delta y=2x(\delta x)+(\delta x)^2$$

So the gradient of PQ is,

$$\frac{\delta y}{\delta x}=\frac{2x(\delta x)+(\delta x)^2}{\delta x}$$
$$=\frac{\delta x(2x+\delta x)}{\delta x}$$
$$=2x+\delta x$$

As we let dx become zero we are left with just 2x, and this is the formula for the gradient of the tangent at P. We have a concise way of expressing the fact that we are letting dx approach zero. We write,

Gradient of tangent = $\lim\limits_{\delta x\to o}\frac{\delta y}{\delta x}=\lim\limits_{\delta x\to o}(2x+\delta x)=2x$

'lim' stands for 'limit 'and we say that the limit, as x tends to zero, of 2x+dx is 2x. Note that when x has the value 3, 2x has the value 6, and so this general result agrees with the earlier result when we calculated the gradient at the point P(3, 9).

We can do this calculation in the same way for lots of curves. We have a special symbol for the phrase,

$$\lim_{\delta x\to o}\frac{\delta y}{\delta x}$$

We write this as dy/dx and say this as "dee y by dee x". This is also referred to as the derivative of y with respect to x.

Use of Function Notation

We often use function notation y = f(x). Then, the point P has coordinates (x, f(x)). Point Q has coordinates (x + dx, f(x + dx)). So, the change in y, that is dy is f(x + dx) – f(x). Then,

$$\frac{\delta y}{\delta x}=\frac{f(x+\delta x)-f(x)}{\delta x}$$

So,

$$\frac{dy}{dx}=\lim_{\delta x\to o}\frac{\delta y}{\delta x}$$
$$=\lim_{\delta x\to o}\frac{f(x+\delta x)-f(x)}{\delta x}$$

This is the definition, for any function y = f(x), of the derivative, dy/dx. Note: Given y = f(x), its derivative, or rate of change of y with respect to x is defined as,

$$\frac{dy}{dx} = \lim_{\delta x \to 0} \frac{f(x+\delta x)-f(x)}{\delta x}$$

Example

Suppose we want to differentiate the function f(x) = 1/x from first principles. We have marked point P(x, f(x)) and the neighboring point Q(x + dx, f(x +d x)).

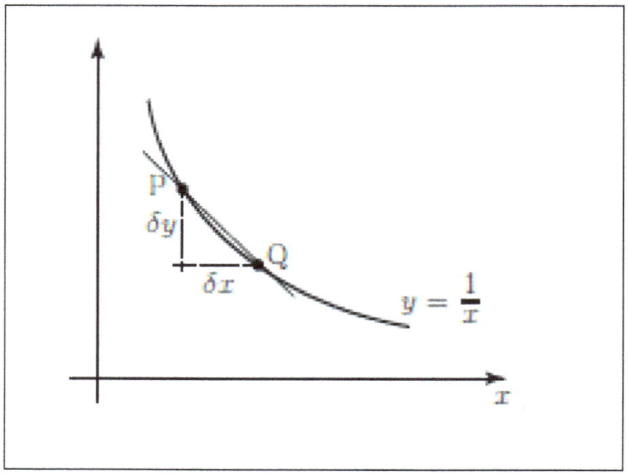

$$\frac{\delta y}{\delta x} = \frac{f(x+\delta x)-f(x)}{\delta x}$$

$$= \frac{\frac{1}{x+\delta x} - \frac{1}{x}}{\delta x}$$

$$= \frac{\frac{x-(x+\delta x)}{x(x+\delta x)}}{\delta x}$$

$$= \frac{-\delta x}{x(x+\delta x)\delta x}$$

$$= \frac{-1}{(x+\delta x)x}$$

In the limit as $\delta x \to 0$ this becomes,

$$\frac{dy}{dx} = -\frac{1}{x^2}$$

So the derivative of $y = \frac{1}{x}$ with respect to x is $-\frac{1}{x^2}$.

Rules of Differentiation

We have seen how to find the derivative of a function using the definition as it is too tedious and time-consuming. We want to develop formulas to help us find derivatives more easily.

Constant Rule

If $f(x) = c$, where c is a constant, then $f'(x) = 0$. In other words, the derivative of a constant is zero.

E.g. if $y = 2$, then $\frac{dy}{dx} = 0$.

Derivative of x to a power:

Power Rule

If $f(x) = x^n$, where n is any real number, then $f'(x) = n^{xn-1}$. If we use the prime notation, understanding that we are differentiating with respect to x, we have:

$$\left(x^n\right)' = nx^{n-1}$$

E.g. $y = x^7$, then $\frac{dy}{dx} = 7x^6$. In applying the power rule, n can be any real number. Consider $f(x) = x^{\frac{2}{3}}$, then,

$$f'(x) = \frac{2}{3}x^{\frac{2}{3}-1} = \frac{2}{3}x^{-\frac{1}{3}} = \frac{2}{3\sqrt[3]{x}}$$

$f(x) = x^{\sqrt{3}}$, then,

$$f'(x) = \sqrt{3x}^{\sqrt{3}-1}$$

$f(x) = x^{\pi}$, then,

$$f'(x) = \pi x^{\pi-1}$$

Constant Multiple Rule

If c is a constant, $g(x) = cf(x)$, and f is differentiable, then g is differentiable and $g'(x) = cf'(x)$. i.e. the derivative of a constant times a function is the constant times the derivative of the function.

$$(cf(x))' = cf'(x)$$

E.g. $y = 3x^5$, then $\frac{dy}{dx} = 3(x^5)' = 3(5x^4) = 15x^4$.

Sum and Difference Rule

Suppose h(x) = f(x) + g(x), and f and g are both differentiable, then h is also differentiable and h' (x) = f' (x) + g' (x). i.e. The derivative of the sum is the sum of the derivatives.

$$[f(x) + g(x)]' = f'(x) + g'(x)$$

Suppose h(x) = f(x) – g(x), and f and g are both differentiable, then h is also differentiable and h' (x) = f' (x) – g' (x). i.e. The derivative of the difference is the difference of the derivatives.

$$[f(x) - g(x)]' = f'(x) - g'(x)$$

Example: We may combine the rules to find complicated derivatives such as this:

$$f(x) = \sqrt{x} + x^3 - 4x^2 + 2x - 5$$

To find f'(x), let us rewrite the radical in a more convenient form:

$$f(x) = x^{1/2} + x^3 - 4x^2 + 2x - 5$$

Using the sum and different rule, we know that,

$$\begin{aligned}\frac{dy}{dx} &= \left(x^{1/2}\right)' + \left(x^3\right)' - \left(4x^2\right)' + (2x)' - (5)' \\ &= \frac{1}{2}x^{(1/2)-1} + 3x^2 - 4\left(x^2\right)' + 2(x)' \\ &= \frac{1}{2}x^{-1/2} + 3x^2 - 4(2x) + 2 \\ &= \frac{1}{2\sqrt{x}} + 3x^2 - 8x + 2\end{aligned}$$

Product Rule

Suppose h(x) = f(x) · g(x) and f and g are both differentiable, then h is differentiable, and h' (x) = g(x) · f' (x) + f(x) · g' (x) .i.e.

$$\left[f(x) \cdot g(x)\right]' = g(x)f'(x) + f(x)g'(x)$$

Example: Find the derivative of:

$$y = \left(x^4 + 2x^2 - x\right)\left(2x^7 - 9x^4 - 3x + 4\right)$$

Notice that y is a product of two functions, (x^4+2x^2-x) and $(2x^7-9x^4-3x+4)$. To find $\frac{dy}{dx}$ we use the product rule:

$$\frac{dy}{dx} = \left(x^4 + 2x^2 - x\right)\left(2x^7 - 9x^4 - 3x + 4\right)' + \left(2x^7 - 9x^4 - 3x + 4\right)\left(x^4 + 2x^2 - x\right)'$$

$$=(x^4+2x^2-x)(14x^6-36x^3-3)+(2x^7-9x^4-3x+4)(4x^3+4x-1)$$

It is important to note that,

$$[f(x)\cdot g(x)]' \neq f'(x)\cdot g'(x)$$

In other words, the derivative of a product is not the product of the derivatives.

Quotient Rule

Suppose $h(x)=\dfrac{f(x)}{g(x)}$, and f and g are differentiable, then h is differentiable where $g(x)\neq 0$ and

$h'(x)=\dfrac{g(x)\cdot f'(x)-f(x)\cdot g'(x)}{[g(x)]^2}$. i.e.,

$$\left[\frac{f(x)}{g(x)}\right]'=\frac{g(x)f'(x)-f(x)g'(x)}{[g(x)]^2}$$

$$\left[\frac{f(x)}{g(x)}\right]'\neq\frac{f'(x)}{g'(x)}$$

In other words, the derivative of a quotient is not the quotient of the derivatives. Another important note is that in the numerator of the quotient rule, we are taking the difference of the two terms, so the order of which derivative to take first matters.

Let, $y=\dfrac{x^3-4x^2+1}{2x^2-x}$, find $\dfrac{dy}{dx}$.

Since this is a quotient, we use the quotient rule:

$$\frac{dy}{dx}=\frac{(2x^2-x)[x^3-4x^2+1]'-(x^3-4x^2+1)[2x^2-x]'}{(2x^2-x)^2}$$

$$=\frac{(2x^2-x)(3x^2-8x)-(x^3-4x^2+1)(4x-1)}{(2x^2-x)^2}$$

Example: Let $y=\dfrac{(\sqrt{x}-3x^4)(x^2+1)}{(2x+1)}$, find $\dfrac{dy}{dx}$.

This is a function which is a quotient, so in order to take its derivative; we must use the quotient rule:

$$\frac{dy}{dx}=\frac{[2x+1]\left[\left(\sqrt{x}-3x^4\right)\left(x^2+1\right)\right]'-\left[\left(\sqrt{x}-3x^4\right)\left(x^2+1\right)\right][2x+1]'}{(2x+1)^2}$$

However, the numerator is itself a product, so to differentiate the numerator; we must use the product rule:

$$\left[\left(\sqrt{x}-3x^4\right)\left(x^2+1\right)\right]'$$

$$=\left(x^2+1\right)\left(\sqrt{x}-3x^4\right)'+\left(\sqrt{x}-3x^4\right)\left(x^2+1\right)'$$

$$=\left(x^2+1\right)\left(\frac{1}{2\sqrt{x}}-12x^3\right)+\left(\sqrt{x}-3x^4\right)(2x)$$

So the derivative of y becomes:

$$\frac{dy}{dx}=\frac{[2x+1]\left[\left(x^2+1\right)\left(\frac{1}{2\sqrt{x}}-12x^3\right)+\left(\sqrt{x}-3x^4\right)(2x)\right]-\left[\left(\sqrt{x}-3x^4\right)\left(x^2+1\right)\right]2}{(2x+1)^2}$$

If we use the prime (') notation, we can summarize the differentiation rules in compact form:

$$(c)'=0$$
$$\left(x^n\right)'=nx^{n-1}$$
$$(cf)'=cf'$$
$$(f+g)'=f'+g'$$
$$(f-g)'=f'-g'$$
$$(fg)'=gf'+fg'$$
$$\left(\frac{f}{g}\right)'=\frac{gf'-fg'}{g^2}$$

All the differentiation formulas introduced so far allows us to find derivatives of power, sum, difference, products and quotients. However, none of the differentiation formula allows us to find composition of functions. For example, suppose we want to find the derivative of $y=\sqrt{x^3+x+2}$, how are we going to do that? In order to differentiate the above function, or any function that are the result of composition of two or more functions, we must use:

Chain Rule

Suppose f and g are differentiable functions and h = f ∘ g is the composite function defined by h(x) = f(g(x)), then h is differentiable and,

$$h'(x) = f'(g(x))(g'(x))$$

If we use the d notation, and let y = f(u) and u = g(x), then we have:

$$\frac{dy}{dx} = \frac{dy}{du} \cdot \frac{du}{dx}$$

The chain rule says that the derivative of the composition of two functions is equal to the derivative of the first (the outside) function evaluated at the second (the inside) function times the derivative of the second (the inside) function.

Example: Find the derivative of $y = (x^4 - 5x^3 + 2x - 1)^8$.

Notice that y is really a composition of two functions. If we were to differentiate y (with respect to x), we need to use the chain rule. If we want to use the chain rule formaly, here's how it is done:

Let $f(x) = x^8$, $g(x) = x^4 - 5x^3 + 2x - 1$, then $h(x) = (x^4 - 5x^3 + 2x - 1)^8 = f(g(x))$ is the function we want to differentiate. The chain rule says that we should differentiate the first (outside) function, f, then evaluate it on the second function, g. i.e $h'(x) = f'(g(x))\, g'(x)$.

Since $f'(x) = 8x^7$, and $g(x) = x^4 - 5x^3+2x - 1$, so we have,

$F'(g(x)) = 8(x^4 - 5x^3 + 2x - 1)^7$

Now, $g'(x) = 4x^3 - 15x^2 + 2$, so according to the chain rule,

$h'(x) = f'(g(x))g'(x) = 8(x^4 - 5x^3 + 2x - 1)^7 (4x^3 - 15x^2 + 2)$

Another way to apply the chain rule, using $\frac{dy}{dx} = \frac{dy}{du} \cdot \frac{du}{dx}$.

We want to differentiate $y = (x^4 - 5x^3 + 2x - 1)^8$

Let $u = x^4 - 5x^3 + 2x - 1$ (u is the inside), then $y = u^8$. We want to find $\frac{dy}{dx}$. According to the chain rule,

$$\frac{dy}{dx} = \frac{dy}{du} \cdot \frac{du}{dx}$$

Since $\frac{dy}{du} = 8u^7$, and $\frac{du}{dx} = 4x^3 - 15x^2 + 2$, so,

$$\frac{dy}{dx} = (8u^7)(4x^3 - 15x^2 + 2) = 8(x^4 - 5x^3 + 2x - 1)^7 (4x^3 - 15x^2 + 2)$$

The above method of applying the chain rule is the proper way to use the chain rule. However, most of the time people find it more convenient (and less confusing) to think of the chain rule as derivative of the outside followed by derivative of the inside. For the above example, $y = (x^4 - 5x^3 + 2x - 1)^8$, the outside function is the eight power function, $()^8$; the inside function is $x^4 - 5x^3 + 2x - 1$, so,

$$\frac{dy}{dx} = \frac{d}{dx}(\text{outside})\frac{d}{dx}(\text{inside})$$
$$= \left(x^4 - 5x^3 + 2x - 1\right)^7 \left(4x^3 - 15x^2 + 2\right)$$

It is important to note that, when you differentiate the outside, the inside is unchanged, then you differentiate the inside, and multiply the two. Never try to differentiate the outside and the inside at the same time. In the above example, $y = (x^4 - 5x^3 + 2x - 1)^8$,

$$\frac{dy}{dx} \neq 8\left(4x^3 - 15x^2 + 2\right)^7$$

Example: Let $y = \sqrt{x^3 + x + 2}$, find $\frac{dy}{dx}$.

This is once again a composition of two functions. The outside function is the squareroot √ function, and the inside is the $x^3 + x + 2$ function. The derivative of the outside is $\frac{1}{2\sqrt{}}$, and the derivative of the inside is $3x^2 + 1$.

$$\frac{dy}{dx} = \frac{1}{2\sqrt{x^3 + x + 2}}\left(3x^2 + 1\right) = \frac{3x^2 + 1}{2\sqrt{x^3 + x + 2}}$$

Once again, when you differentiate the outside, do not change the inside. You differentiate the outside with the inside unchanged, then differentiate the inside and multiply the two together.

If you want to differentiation composition of more than two functions, you will need to apply the chain rule multiple times.

Example: Let $f(x) = \sqrt{\left(3x^2 + 5\right)^4 + 1}$, find f'(x).

Notice that f is a composition of three functions. The outer-most squareroot √ function, the middle fourth power plus one $[()^4 + 1]$ function, and the inner-most $3x^2+5$ function. To differentiate this function, we can treat the middle fourth power plus one $()^4 + 1$ function and the inner-most $3x^2 + 5$ function together as one single function, apply the chain rule once; then apply the chain rule once again to differentiate the composition of the middle function with the inner-most function.

$$f'(x) = \frac{1}{2\sqrt{\left(3x^2 + 5\right)^4 + 1}}\left[\left(3x^2 + 5\right)^4 + 1\right]'$$

To differentiate the $(3x^2 + 5)^4 + 1$ function, we have to apply the chain rule one more time:

$$[(3x^2 + 5)^4 + 1]' = [4(3x^2 + 5)^3][6x]$$

So,

$$f'(x) = \frac{1}{2\sqrt{\left(3x^2 + 5\right)^4 + 1}}\left[\left(3x^2 + 5\right)^4 + 1\right]'$$

$$=\frac{1}{2\sqrt{(3x^2+5)^4+1}}\left[4(3x^2+5)^3\right][6x]$$

$$=\frac{24x(3x^2+5)^3}{2\sqrt{(3x^2+5)^4+1}}=\frac{12x(3x^2+5)^3}{\sqrt{(3x^2+5)^4+1}}$$

Let's do a problem that incorporates all the rules that we've learned so far:

Let $f(x)=\frac{(3x+1)^8(x-1)}{x^2-4x}$, find f' (x).

This function is a quotient, so the first thing we want to do is to apply the quotient rule:

$$f'(x)=\frac{(x^2-4x)\left[(3x+1)^8(x-1)\right]'-\left[(3x+1)^8(x-1)\right]\left[x^2-4x\right]'}{(x^2-4x)^2}$$

Since the numerator is a product, we must use the product rule to differentiate the numerator.

$$\left[(3x+1)^8(x-1)\right]'=(x-1)\left[(3x+1)^8\right]'+\left[(3x+1)^8\right](x-1)'$$

To differentiate the first factor, which involves an outside $[()^8]$ and an inside 3x + 1, we must apply the chain rule, this gives us:

$$\left[(3x+1)^8(x-1)\right]'=(x-1)\left[(3x+1)^8\right]'+(3x+1)^8(x-1)'$$

$$=(x-1)\left[8(3x+1)^7(3)\right]+(3x+1)^8(1)$$

$$=24(x-1)(3x+1)^7+(3x+1)^8$$

The derivative of the function $f(x)=\frac{(3x+1)^8(x-1)}{x^2-4x}$ can now be completed:

$$f'(x)=\frac{(x^2-4x)\left[(3x+1)^8(x-1)\right]'-\left[(3x+1)^8(x-1)\right]\left[x^2-4x\right]'}{(x^2-4x)^2}$$

$$=\frac{(x^2-4x)\left[24(x-1)(3x+1)^7+(3x+1)^8\right]-\left[(3x+1)^8(x-1)\right][2x-4]}{(x^2-4x)^2}$$

Differentiation of Trigonometric Functions

Here, we look at how to differentiate the functions f(x) = sin x and f(x) = cos x from first principles. We need to remind ourselves of some familiar results.

1. The derivative of f(x). The definition of the derivative of a function y = f(x) is,

$$\frac{dy}{dx} = \lim_{\delta x \to 0} \frac{f(x+\delta x) - f(x)}{\delta x}$$

2. Two trigonometric identities. We will make use of the trigonometric identities:

$$\sin C - \sin D = 2\cos\frac{C+D}{2}\sin\frac{C-D}{2}$$

$$\cos C - \cos D = -2\sin\left(\frac{C+D}{2}\right)\sin\left(\frac{C-D}{2}\right)$$

3. The limit of the function $\frac{\sin\theta}{\theta}$. As θ (measured in radians) approaches zero, the function $\frac{\sin\theta}{\theta}$ tends to 1. We write this as:

$$\lim_{\theta \to 0} \frac{\sin\theta}{\theta} = 1$$

This result can be justified by choosing values of θ closer and closer to zero and examining the behaviour of $\frac{\sin\theta}{\theta}$. Table below shows values of θ and $\frac{\sin\theta}{\theta}$ as θ becomes smaller.

Table: The value of $\frac{\sin\theta}{\theta}$ as θ tends to zero is 1.

θ	sinθ	$\frac{\sin\theta}{\theta}$
1	0.84147	0.84147
0.1	0.09983	0.99833
0.01	0.00999	0.99983

You should verify these results with your calculator to appreciate that the value of $\frac{\sin\theta}{\theta}$ approaches 1 as θ tends to zero. We now use these results in order to differentiate f(x) = sin x from first principles.

Differentiating f(x) = sin x

Here f(x) = sin x so that f(x + δx) = sin(x + δx). So,

$$f(x + \delta x) - f(x) = \sin(x + \delta x) - \sin x$$

The right hand side is the difference of two sine terms. We use the first trigonometric identity (above) to write this in an alternative form.

$$\sin(x+\delta x)-\sin x = 2\cos\frac{x+\delta x+x}{2}\sin\frac{\delta x}{2}$$
$$= 2\cos\frac{2x+\delta x}{2}\sin\frac{\delta x}{2}$$
$$= 2\cos\left(x+\frac{\delta x}{2}\right)\sin\frac{\delta x}{2}$$

Then, using the definition of the derivative:

$$\frac{dy}{dx} = \lim_{\delta x\to 0}\frac{f(x+\delta x)-f(x)}{\delta x}$$
$$= \frac{2\cos\left(x+\frac{\delta x}{2}\right)\sin\frac{\delta x}{2}}{\delta x}$$

The factor of 2 can be moved into the denominator as follows, in order to write this in an alternative form:

$$\frac{dy}{dx} = \frac{\cos\left(x+\frac{\delta x}{2}\right)\sin\frac{\delta x}{2}}{\delta x/2}$$
$$= \cos\left(x+\frac{\delta x}{2}\right)\frac{\sin\frac{\delta x}{2}}{\frac{\delta x}{2}}$$

We now let δx tend to zero. Consider the term $\frac{\sin\frac{\delta x}{2}}{\frac{\delta x}{2}}$ and use the result that $\lim_{\theta\to 0}\frac{\sin\theta}{\theta} = 1$ with $\theta = \frac{\delta x}{2}$. We see that:

$$\lim_{\delta x\to 0}\frac{\sin\frac{\delta x}{2}}{\frac{\delta x}{2}} = 1$$

Further,

$$\lim_{\delta x\to 0}\cos\left(x+\frac{\delta x}{2}\right) = \cos x$$

So finally,

$$\frac{dy}{dx} = \cos x$$

The Derivative of f(x) = cos x

Here $f(x) = \cos x$ so that $f(x + \delta x) = \cos(x + \delta x)$. So,

$$f(x + \delta x) - f(x) = \cos(x + \delta x) - \cos x$$

The right hand side is the difference of two cosine terms. This time we use the trigonometric identity:

$$\cos C - \cos D = -2\sin\left(\frac{C+D}{2}\right)\sin\left(\frac{C-D}{2}\right)$$

to write this in an alternative form.

$$\cos(x+\delta x) - \cos x = -2\sin\frac{x+\delta x+x}{2}\sin\frac{\delta x}{2}$$

$$-2\sin\frac{2x+\delta x}{2}\sin\frac{\delta x}{2}$$

$$= -2\sin\left(x+\frac{\delta x}{2}\right)\sin\frac{\delta x}{2}$$

Then, using the definition of the derivative:

$$\frac{dy}{dx} = \lim_{\delta x \to 0}\frac{f(x+\delta x)-f(x)}{\delta x}$$

$$= \frac{-2\sin\left(x+\frac{\delta x}{2}\right)\sin\frac{\delta x}{2}}{\delta x}$$

The factor of 2 can be moved as before, in order to write this in an alternative form:

$$\frac{dy}{dx} = -\frac{\sin\left(x+\frac{\delta x}{2}\right)\sin\frac{\delta x}{2}}{\delta x/2}$$

$$= -\sin\left(x+\frac{\delta x}{2}\right)\frac{\sin\frac{\delta x}{2}}{\frac{\delta x}{2}}$$

We now want to let δx tend to zero. As before:

$$\lim_{\delta x \to 0}\frac{\sin\frac{\delta x}{2}}{\frac{\delta x}{2}} = 1$$

Further,

$$\lim_{\delta x \to 0} -\sin\left(x+\frac{\delta x}{2}\right) = -\sin x$$

So finally,

$$\frac{dy}{dx} = -\sin x$$

So, we have used differentiation from first principles to find the derivatives of the functions sin x and cos x.

Derivatives of Exponential and Logarithmic Function

Here, we explain how to differentiate the functions in x and e^x from first principles. To understand what follows we need to use the result that the exponential constant e is defined as the limit as t tends to zero of $(1 + t)^{1/t}$ i.e. $\lim_{t\to 0} (1 + t)^{1/t}$.

We have evaluated the expression $(1 + t)^{1/t}$ for a number of decreasing values of t as shown in Table below. Note that as t gets closer to zero, the value of the expression gets closer to the value of the exponential constant e≈ 2.718.

t	$(1+t)^{1/t}$	
1	$(1+1)^{1/1}$	=2
0.1	$(1+0.1)^{1/0.1}$	=2.594
0.01	$(1+0.01)^{1/0.01}$	=2.705
0.001	$(1.001)^{1/0.001}$	2.717
0.0001	$(1.0001)^{1/0.0001}$	2.718

We will also make frequent use of the laws of indices and the laws of logarithms.

Differentiation of a Function f(x)

To differentiate any function, f(x), from first principles we find the slope, $\frac{\delta x}{\delta y}$, of the line joining an arbitrary point, A, and a neighbouring point, B, on the graph of f(x). We then determine what happens to $\frac{\delta x}{\delta y}$ in the limit as δx tends to zero.

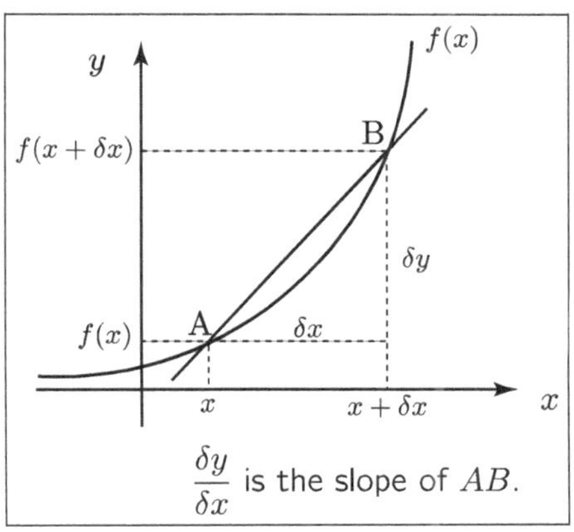

$\frac{\delta x}{\delta y}$ is the slope of AB.

The derivative, f′ (x), is then given by,

$$f'(x) = \lim_{\delta x \to 0} \frac{\delta y}{\delta x} = \lim_{\delta x \to 0} \frac{f(x+\delta x) - f(x)}{\delta x}$$

Differentiation of f(x) = ln x

Using the definition of the derivative in the case when f(x) = ln x we find:

$$\frac{f(x+\delta x) - f(x)}{\delta x} = \frac{\ln(x+\delta x) - \ln x}{\delta x}$$

We proceed by using the law of logarithms $\log A - \log B = \log \frac{A}{B}$ to re-write the right-hand side as firstly:

$$\frac{1}{\delta x}\left(\ln(x+\delta x) - \ln x\right) = \frac{1}{\delta x}\ln\left(\frac{x+\delta x}{x}\right) = \frac{1}{\delta x}\ln\left(1 + \frac{\delta x}{x}\right)$$

In order to simplify what will follow we make a substitution: Let $t = \frac{\delta x}{x}$, that is, δx = xt. (This substitution is made because in the calculations which follow it is the ratio of δx to x which turns out to be important. We need not worry about x being zero because we are interested in differentiating ln x and the logarithm function is only defined for positive values of x). Then,

$$\frac{f(x+\delta x) - f(x)}{\delta x} = \frac{1}{xt}\ln(1+t)$$

Further, using the law n log A = log A^n we can take the $\frac{1}{t}$ inside the logarithm to give:

$$\frac{f(x+\delta x) - f(x)}{\delta x} = \frac{1}{x}\ln(1+t)^{\frac{1}{t}}$$

Referring to the general case in Figure, this represents the slope of the line joining the two points on the graph of f(x). To find the derivative we need to let δx tend to zero. Because we substituted $t = \frac{\delta x}{x}$ we need to let t tend to zero.

We have,

$$f'(x) = \lim_{t \to 0} \frac{1}{x}\ln(1+t)^{\frac{1}{t}}$$

In this limiting process it is t which tends to zero, and we can regard x as a fixed number. So, it can be taken outside the limit to give:

$$f'(x) = \frac{1}{x}\lim_{t \to 0}\ln(1+t)^{\frac{1}{t}}$$

But we know that,

$$\lim_{t \to 0}(1+t)^{\frac{1}{t}} = e$$

and so,

$$f'(x) = \frac{1}{x}\ln e = \frac{1}{x}$$

since ln e = 1.

We have shown, from first principles, that the derivative of ln x is equal to $\frac{1}{x}$.

Differentiation of f(x) = e^x^

To differentiate $y = e^x$ we will rewrite this expression in its alternative form using logarithms:

$$\ln y = x$$

Then differentiating both sides with respect to x,

$$\frac{d}{dx}(\ln y) = 1$$

The idea is now to find $\frac{dy}{dx}$.

Recall that $\frac{d}{dx}(\ln y) = \frac{d}{dx}(\ln y) \times \frac{dy}{dx}$. (This result is obtained using a technique known as the chain rule). Now we know, that $\frac{d}{dy}(\ln y) = \frac{1}{y}$ and so,

$$\frac{1}{y}\frac{dy}{dx} = 1$$

Rearranging,

$$\frac{dy}{dx} = y$$

But $y = e^x$ and so we have the important and well-known result that,

$$\frac{dy}{dx} = e^x$$

The exponential function (and multiples of it) is the only function which is equal to its derivative.

Tangent Lines and Rate of Change

The first problem that we're going to take a look at is the tangent line problem. Before getting into this problem it would probably be best to define a tangent line. A tangent line to the function f(x) at the point x=a is a line that just touches the graph of the function at the point in question and is "parallel" (in some way) to the graph at that point.

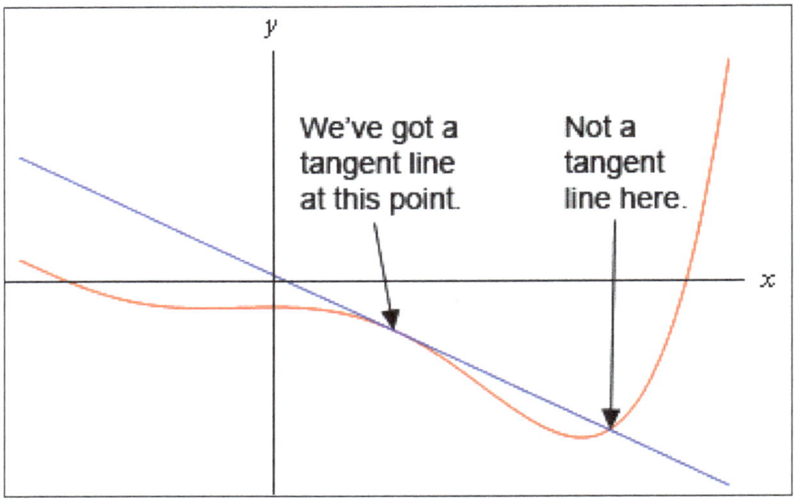

In this graph the line is a tangent line at the indicated point because it just touches the graph at that point and is also "parallel" to the graph at that point. Likewise, at the second point shown, the line does just touch the graph at that point, but it is not "parallel" to the graph at that point and so it's not a tangent line to the graph at that point.

At the second point shown (the point where the line isn't a tangent line) we will sometimes call the line a secant line.

In general, we will think of a line and a graph as being parallel at a point if they are both moving in the same direction at that point. So, in the first point above the graph and the line are moving in the same direction and so we will say they are parallel at that point. At the second point, on the other hand, the line and the graph are not moving in the same direction so they aren't parallel at that point.

There are a couple of important points to note about our work above. First, we looked at points that were on both sides of x=1. In this kind of process it is important to never assume that what is happening on one side of a point will also be happening on the other side as well. We should always look at what is happening on both sides of the point. In this example we could sketch a graph and from that guess that what is happening on one side will also be happening on the other, but we will usually not have the graphs in front of us or be able to easily get them.

Next, notice that when we say we're going to move in close to the point in question we do mean that we're going to move in very close and we also used more than just a couple of points. We should never try to determine a trend based on a couple of points that aren't really all that close to the point in question.

The next thing to notice is really a warning more than anything. The values of m_{PQ} in this example were fairly "nice" and it was pretty clear what value they were approaching after a couple of computations. In most cases this will not be the case. Most values will be far "messier" and you'll often need quite a few computations to be able to get an estimate. You should always use at least four points, on each side to get the estimate. Two points is never sufficient to get a good estimate and three points will also often not be sufficient to get a good estimate. Generally, you keeping picking points closer and closer to the point you are looking at until the change in the value between two successive points is getting very small.

Last, we were after something that was happening at x=1 and we couldn't actually plug x=1 into our formula for the slope. Despite this limitation we were able to determine some information about what was happening at x=1 simply by looking at what was happening around x=1.

Before moving on let's do a quick review of just what we did in the above example. We wanted the tangent line to f(x) at a point x=a. First, we know that the point P=(a,f(a)) will be on the tangent line. Next, we'll take a second point that is on the graph of the function, call it Q=(x,f(x)) and compute the slope of the line connecting P and Q as follows,

$$m_{PQ} = \frac{f(x) - f(a)}{x - a}$$

We then take values of x that get closer and closer to x=ax=a (making sure to look at x's on both sides of x=a and use this list of values to estimate the slope of the tangent line, m. The tangent line will then be,

$$y = f(a) + m(x - a)$$

Rates of Change

Here we are going to consider a function, f(x), that represents some quantity that varies as x varies. For instance, maybe f(x) represents the amount of water in a holding tank after x minutes. Or maybe f(x) is the distance traveled by a car after xx hours. In both of these examples we used x to represent time. Of course x doesn't have to represent time, but it makes for examples that are easy to visualize.

What we want to do here is determine just how fast f(x) is changing at some point, say x=a. This is called the instantaneous rate of change or sometimes just rate of change of f(x) at x=a.

As with the tangent line problem all that we're going to be able to do at this point is to estimate the rate of change. So, let's continue with the examples above and think of f(x) as something that is changing in time and x being the time measurement. Again, x doesn't have to represent time but it will make the explanation a little easier. While we can't compute the instantaneous rate of change at this point we can find the average rate of change.

To compute the average rate of change of f(x) at x=a all we need to do is to choose another point, say x, and then the average rate of change will be,

$$\text{A.R.C.} = \frac{\text{change in } f(x)}{\text{change in } x}$$
$$= \frac{f(x)-f(a)}{x-a}$$

Then to estimate the instantaneous rate of change at x=a all we need to do is to choose values of xx getting closer and closer to x=a (don't forget to choose them on both sides of x=a) and compute values of A.R.C. We can then estimate the instantaneous rate of change from that.

So, just what does this tell us about the volume at t=5? Let's put some units on the answer from above. This might help us to see what is happening to the volume at this point. Let's suppose that the units on the volume were in cm^3. The units on the rate of change (both average and instantaneous) are then cm^3/hr.

We have estimated that at t=5 the volume is changing at a rate of 15 cm^3/hr. This means that at t=5 the volume is changing in such a way that, if the rate were constant, then an hour later there would be 15 cm^3 more air in the balloon than there was at t=5.

We do need to be careful here however. In reality there probably won't be 15 cm^3 more air in the balloon after an hour. The rate at which the volume is changing is generally not constant so we can't make any real determination as to what the volume will be in another hour. What we can say is that the volume is increasing, since the instantaneous rate of change is positive, and if we had rates of change for other values of t we could compare the numbers and see if the rate of change is faster or slower at the other points.

For instance, at t=4 the instantaneous rate of change is 0 cm^3/hr and at t=3 the instantaneous rate of change is -9 cm^3/hr. We'll leave it to you to check these rates of change. In fact, that would be a good exercise to see if you can build a table of values that will support our claims on these rates of change.

Anyway, back to the example. At t=4 the rate of change is zero and so at this point in time the volume is not changing at all. That doesn't mean that it will not change in the future. It just means that exactly at t=4 the volume isn't changing. Likewise, at t=3 the volume is decreasing since the rate of change at that point is negative. We can also say that, regardless of the increasing/decreasing aspects of the rate of change, the volume of the balloon is changing faster at t=5 than it is at t=3 since 15 is larger than 9.

Velocity Problem

Let's briefly look at the velocity problem. Many calculus books will treat this as its own problem. We however, like to think of this as a special case of the rate of change problem. In the velocity problem we are given a position function of an object, f(t), that gives the position of an object at time t. Then to compute the instantaneous velocity of the object we just need to recall that the velocity is nothing more than the rate at which the position is changing.

In other words, to estimate the instantaneous velocity we would first compute the average velocity,

$$\text{A.V.} = \frac{\text{change in position}}{\text{time traveled}}$$
$$= \frac{f(t)-f(a)}{t-a}$$

and then take values of t closer and closer to t=a and use these values to estimate the instantaneous velocity.

Change of Notation

There is one last thing that we need to do in this section before we move on. The main point of this section was to introduce us to a couple of key concepts and ideas that we will see throughout the first portion of this course as well as get us started down the path towards limits.

Before we move into limits officially let's go back and do a little work that will relate both (or all three if you include velocity as a separate problem) problems to a more general concept.

First, notice that whether we wanted the tangent line, instantaneous rate of change, or instantaneous velocity each of these came down to using exactly the same formula. Namely,

$$\frac{f(x)-f(a)}{x-a}$$

This should suggest that all three of these problems are then really the same problem. In fact this is the case as we will see in the next chapter. We are really working the same problem in each of these cases the only difference is the interpretation of the results.

In preparation for the next section where we will discuss this in much more detail we need to do a quick change of notation. It's easier to do here since we've already invested a fair amount of time into these problems.

In all of these problems we wanted to determine what was happening at x=a. To do this we chose another value of x and plugged into equation above. For what we were doing here that is probably most intuitive way of doing it. However, when we start looking at these problems as a single problem equation above will not be the best formula to work with.

What we'll do instead is to first determine how far from x=a. we want to move and then define our new point based on that decision. So, if we want to move a distance of h from x=a the new point would be x=a+h. This is shown in the sketch below.

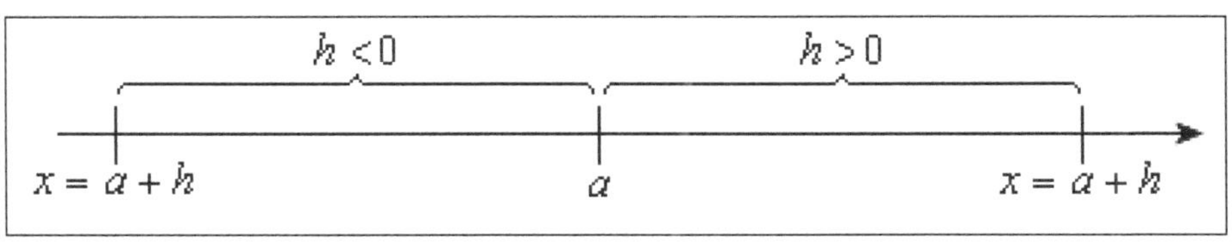

As we saw in our work above it is important to take values of x that are both sides of x=a. This way of choosing new value of x will do this for us as we can see in the sketch above. If h>o we will get value of xx that are to the right of x=ax=a and if h<o we will get values of xx that are to the left of x=a and both are given by x=a+h.

Now, with this new way of getting a second value of x in $\frac{f(x)-f(a)}{x-a}$ will become,

$$\frac{f(x)-f(a)}{x-a}=\frac{f(a+h)-f(a)}{a+h-a}=\frac{f(a+h)-f(a)}{h}$$

Now, this is for a specific value of x, *i.e.* x=a and we'll rarely be looking at these at specific values of x. So, we take the final step in the above equation and replace the a with x to get,

$$\frac{f(x+h)-f(x)}{h}$$

This gives us a formula for a general value of xx and on the surface it might seem that this is going to be an overly complicated way of dealing with this stuff. However, as we will see it will often be easier to deal with this form than it will be to deal with the original form, $\frac{f(x)-f(a)}{x-a}$.

Maxima and Minima

When using mathematics to model the physical world in which we live, we frequently express physical quantities in terms of variables. Then, functions are used to describe the ways in which these variables change. A scientist or engineer will be interested in the ups and downs of a function, its maximum and minimum values, its turning points. Drawing a graph of a function using a graphical calculator or computer graph plotting package will reveal this behaviour, but if we want to know the precise location of such points we need to turn to algebra and differential calculus. In this section we look at how we can find maximum and minimum points in this way. Consider the graph of the function, y(x), shown in Figure below. If, at the points marked A, B and C, we draw tangents to the graph, note that these are parallel to the x axis. They are horizontal. This means that at each of the points A, B and C the gradient of the graph is zero.

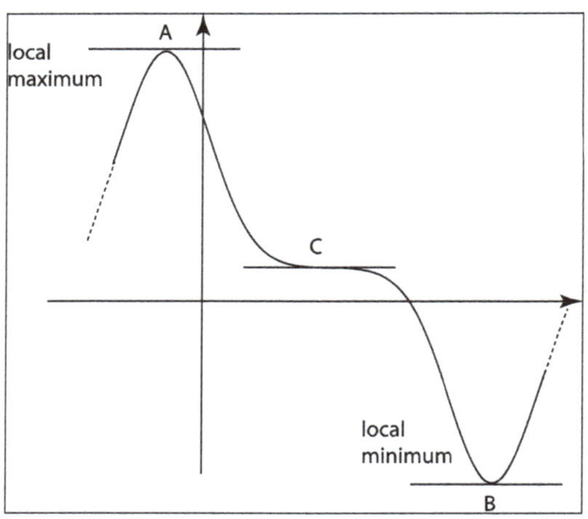

The gradient of this graph is zero at each of the points A, B and C.

We know that the gradient of a graph is given by $\frac{dy}{dx}$ Consequently, $\frac{dy}{dx} = 0$ at points A, B and C. All of these points are known as stationary points.

Turning Points

In above figure, Notice that at points A and B the curve actually turns. These two stationary points are referred to as turning points. Point C is not a turning point because, although the graph is flat for a short time, the curve continues to go down as we look from left to right.

So, all turning points are stationary points. But not all stationary points are turning points (e.g. point C). In other words, there are points for which $\frac{dy}{dx} = 0$ which are not turning points.

Point A in previous figure is called a local maximum because in its immediate area it is the highest point, and so represents the greatest or maximum value of the function. Point B in figure is called a local minimum because in its immediate area it is the lowest point, and so represents the least, or minimum, value of the function. Loosely speaking, we refer to a local maximum as simply a maximum. Similarly, a local minimum is often just called a minimum.

Distinguishing Maximum Points from Minimum Points

Think about what happens to the gradient of the graph as we travel through the minimum turning point, from left to right, that is as x increases. Study figure below to help you do this.

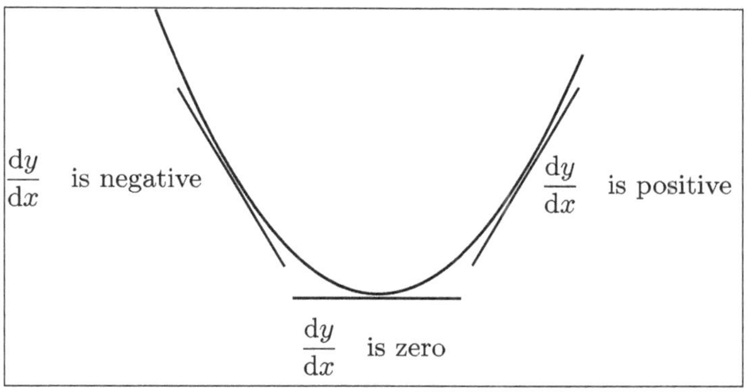

$\frac{dy}{dx}$ goes from negative through zero to positive as x increases.

Notice that to the left of the minimum point, $\frac{dy}{dx}$ is negative because the tangent has negative gradient. At the minimum point, $\frac{dy}{dx} = 0$. To the right of the minimum point $\frac{dy}{dx}$ is positive, because here the tangent has a positive gradient. So, $\frac{dy}{dx}$ goes from negative, to zero, to positive as x increases. In other words, $\frac{dy}{dx}$ must be increasing as x increases. In fact, we can use this observation, once we have found a stationary point, to check if the point is a minimum. If $\frac{dy}{dx}$ is increasing near the stationary point then that point must be minimum. Now, if the derivative of $\frac{dy}{dx}$ is positive then we will know that $\frac{dy}{dx}$ is increasing; so we will know that the stationary point is a minimum.

Now the derivative of $\frac{dy}{dx}$, called the second derivative, is written $\frac{d^2y}{dx^2}$. We conclude that if $\frac{d^2y}{dx^2}$ is positive at a stationary point, then that point must be a minimum turning point.

It is important to realise that this test for a minimum is not conclusive. It is possible for a stationary point to be a minimum even if $\frac{d^2y}{dx^2}$ equals 0, although we cannot be certain: Other types of behaviour are possible. (However, we cannot have a minimum if $\frac{d^2y}{dx^2}$ is negative). To see this consider the example of the function $y = x^4$. A graph of this function is shown in figure below. There is clearly a minimum point when $x = 0$. But $\frac{dy}{dx} = 4x^3$ and this is clearly zero when $x = 0$. Differentiating again $\frac{d^2y}{dx^2} = 12x^2$ which is also zero when $x = 0$.

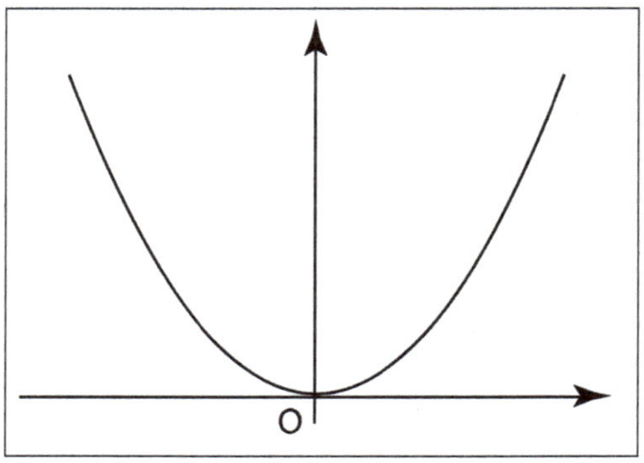

The function $y = x^4$ has a minimum at the origin where $x = 0$, but $\frac{d^2y}{dx^2} = 0$ and so is not greater than 0.

Now think about what happens to the gradient of the graph as we travel through the maximum turning point, from left to right, that is as x increases.

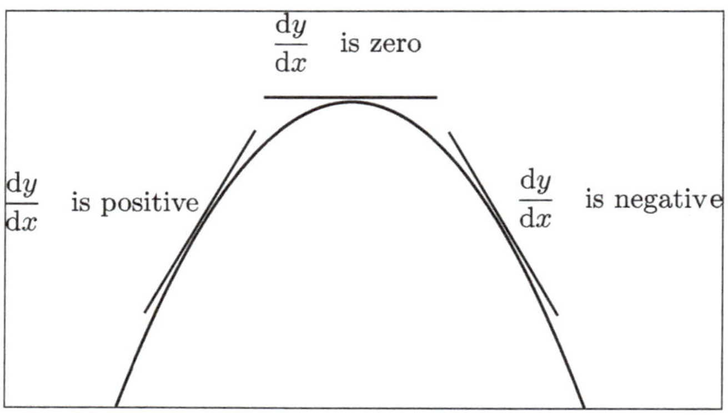

$\frac{dy}{dx}$ from positive through zero to negative as x increases.

Notice that to the left of the maximum point, $\frac{dy}{dx}$ is positive because the tangent has positive gradient. At the maximum point, $\frac{dy}{dx} = 0$. To the right of the maximum point $\frac{dy}{dx}$ is negative, because here the tangent has a negative gradient. So, $\frac{dy}{dx}$ goes from positive, to zero, to negative as x increases.

In fact, we can use this observation to check if a stationary point is a maximum. If $\frac{dy}{dx}$ is decreasing near a stationary point then that point must be maximum. Now, if the derivative of $\frac{dy}{dx}$ is negative then we will know that $\frac{dy}{dx}$ is decreasing; so we will know that the stationary point is a maximum. As before, the derivative of $\frac{dy}{dx}$, the second derivative is $\frac{d^2y}{dx^2}$. We conclude that if $\frac{d^2y}{dx^2}$ is negative at a stationary point, then that point must be a maximum turning point.

It is important to realise that this test for a maximum is not conclusive. It is possible for a stationary point to be a maximum even if $\frac{d^2y}{dx^2} = 0$, although we cannot be certain: other types of behaviour are possible. But we cannot have a maximum if $\frac{d^2y}{dx^2} > 0$, because, as we have already seen the point would be a minimum.

Example: Suppose we wish to find the turning points of the function $y = x^3 - 3x + 2$ and distinguish between them. We need to find where the turning points are, and whether we have maximum or minimum points. First of all we carry out the differentiation and set $\frac{dy}{dx}$ equal to zero. This will enable us to look for any stationary points, including any turning points.

$$y = x^3 - 3x + 2$$

$$\frac{dy}{dx} = 3x^2 - 3$$

At stationary points, $\frac{dy}{dx} = 0$ and so,

$$3x^2 - 3 = 0$$

$$3(x^2 - 1) = 0 \quad \text{(factorising)}$$

$$3(x-1)(x+1) = 0 \quad \text{(factorising the difference of two squares)}$$

It follows that either $x - 1 = 0$ or $x + 1 = 0$ and so either $x = 1$ or $x = -1$.

We have found the x coordinates of the points on the graph where $\frac{dy}{dx} = 0$, that is the stationary points. We need the y coordinates which are found by substituting the x values in the original function $y = x^3 - 3x + 2$.

- When $x = 1$: $y = 1^3 - 3(1) + 2 = 0$.
- When $x = -1$: $y = (-1)^3 - 3(-1) + 2 = 4$.

To summarise, we have located two stationary points and these occur at (1, 0) and (−1, 4). Next we need to determine whether we have maximum or minimum points, or possibly points such as C in Figure which are neither maxima nor minima.

We have seen that the first derivative $\frac{dy}{dx} = 3x^2-3$. Differentiating this we can find the second derivative:

$$\frac{d^2y}{dx^2} = 6x$$

We now take each point in turn and use our test.

- When $x = 1$: $\frac{d^2y}{dx^2} = 6x = 6(1) = 6$. We are not really interested in this value. What is important is its sign. Because it is positive we know we are dealing with a minimum point.
- When $x = -1$: $\frac{d^2y}{dx^2} = 6x = 6(-1) = -6$. Again, what is important is its sign. Because it is negative we have a maximum point.

Finally, to finish this off we produce a quick sketch of the function now that we know the precise locations of its two turning points.

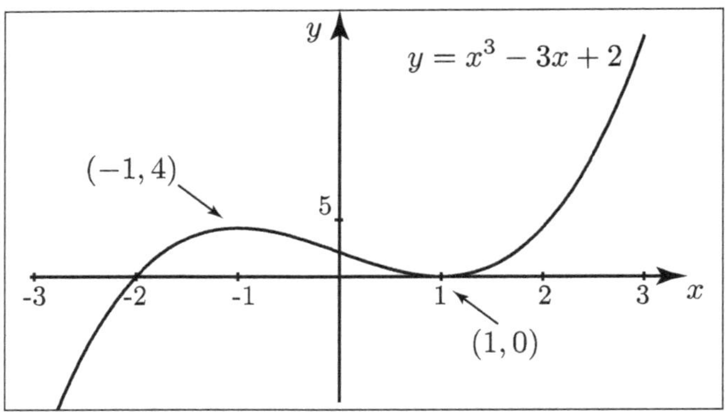

Graph of $y = x^3 - 3x + 2$ showing the turning points.

An Example which uses the First Derivative to Distinguish Maxima and Minima

Suppose we wish to find the turning points of the function $y = \frac{(x-1)^2}{x}$ and distinguish between them. First of all we need to find $\frac{dy}{dx}$.

In this case we need to apply the quotient rule for differentiation.

$$\frac{dy}{dx} = \frac{x \cdot 2(x-1) - (x-1)^2 \cdot 1}{x^2}$$

Look for common factors, and tidy up the expression.

$$\frac{dy}{dx} = \frac{x \cdot 2(x-1) - (x-1)^2 \cdot 1}{x^2}$$

$$= \frac{(x-1)(2x-(x-1))}{x^2}$$

$$= \frac{(x-1)(x+1)}{x^2}$$

We now set $\frac{dy}{dx}$ equal to zero in order to locate the stationary points including any turning points.

$$\frac{(x-1)(x+1)}{x^2} = 0$$

When equating a fraction to zero, it is the top line, the numerator, which must equal zero. Therefore,

$$(x - 1)(x + 1) = 0$$

from which $x - 1 = 0$ or $x + 1 = 0$, and from these equations we find that $x = 1$ or $x = -1$. The y co-ordinates of the stationary points are found from $y = \frac{(x-1)^2}{x}$.

- When $x = 1$: $y = 0$.
- When $x = -1$: $y = \frac{(-2)^2}{-1} = -4$.

We conclude that stationary points occur at $(1, 0)$ and $(-1, -4)$.

We now have to decide whether these are maximum points or minimum points. We could calculate $\frac{d^2y}{dx^2}$ and use the second derivative test as in the previous example. This would involve differentiating $\frac{(x-1)(x+1)}{x^2}$ which is possible but perhaps rather fearsome! Is there an alternative way? The answer is yes. We can look at how $\frac{dy}{dx}$ changes as we move through the stationary point. In essence, we can find out what happens to $\frac{d^2y}{dx^2}$ without actually calculating it.

First consider the point at $x = -1$. We look at what is happening a little bit before the point where $x = -1$, and a little bit afterwards. Often we express the idea of 'a little bit before' and 'a little bit afterwards' in the following way. We can write $-1 - \epsilon$ to represent a little bit less than -1, and $-1 + \epsilon$ to represent a little bit more. The symbol ϵ is the Greek letter epsilon. It represents a small positive quantity, say 0.1. Then $-1 - \epsilon$ would be -1.1, just a little less than -1. Similarly $-1 + \epsilon$ would be -0.9, just a little more than -1.

We now have a look at $\frac{dy}{dx}$; not its value, but its sign.

When $x = -1 - \epsilon$, say -1.1, $\frac{dy}{dx}$ is positive.

When $x = -1$ we already know that $\frac{dy}{dx} = 0$.

When $x = -1 + \epsilon$, say -0.9, $\frac{dy}{dx}$ is negative.

	$x=-1-\epsilon$	$x=-1$	$x=-1+\epsilon$
sign of $\frac{dy}{dx}$	$+$	0	$-$
shape of graph	$\nearrow$	$\longrightarrow$	$\searrow$

Behaviour of the graph near the point (−1, −4).

Figure above shows us that the stationary point at (−1, −4) is a maximum turning point. Then we turn to the point (1, 0). We carry out a similar analysis, looking at the sign of $\frac{dy}{dx}$ at $x = 1 - \epsilon$, $x = 1$, and $x = 1 + \epsilon$. The results are summarised in figure below.

	$x=1-\epsilon$	$x=1$	$x=1+\epsilon$
sign of $\frac{dy}{dx}$	$-$	0	$+$
shape of graph	$\searrow$	$\longrightarrow$	$\nearrow$

Behaviour of the graph near the point (1, 0).

We see that the point is a minimum. This, so-called first derivative test, is also the way to do it if $\frac{d^2y}{dx^2}$ is zero in which case the second derivative test does not work. Finally, for completeness a graph of $y = \frac{(x-1)^2}{x}$ is shown in figure below where you can see the maximum and minimum points.

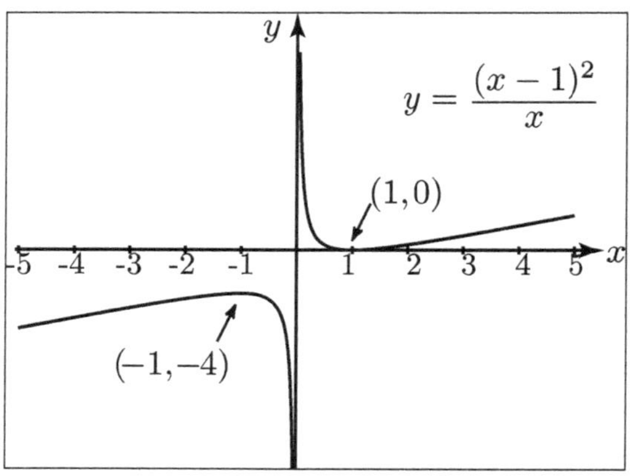

A graph of $y = \frac{(x-1)^2}{x}$ showing the turning points.

References

- Introduction-to-differential-calculus, mathematics-learning-centre: sydney.edu.au, Retrieved 15, April 2020
- Differentiation-first-principles, calculus, pure-maths, advanced-level-maths-revision: revisionmaths.com, Retrieved 07, July 2020
- Calculus-note-intro-derivative: berkeleycitycollege.edu, Retrieved 25, February 2020
- Sin-cos-first-principles: sheffield.ac.uk, Retrieved 30, April 2020
- Mc-ty-logexp-2009: ac.uk, Retrieved 10, August 2020
- Tangents-rates: math.lamar.edu, Retrieved 18, May 2020

3

Higher Order Derivatives

The process of differentiation can be used repeatedly, resulting in the second derivative of the function. It generally has a useful physical interpretation. This chapter closely examines the key concepts of second order derivatives and higher order derivatives as well as Leibniz formula to provide an extensive understanding of the subject.

A branch of mathematics dealing with the concepts of derivative and differential and the manner of using them in the study of functions. The development of differential calculus is closely connected with that of integral calculus. Indissoluble is also their content. Together they form the base of mathematical analysis, which is extremely important in the natural sciences and in technology. The introduction of variable magnitudes into mathematics by R. Descartes was the principal factor in the creation of differential calculus. Differential and integral calculus were created, in general terms, by I. Newton and G. Leibniz towards the end of the 17th century, but their justification by the concept of limit was only developed in the work of A.L. Cauchy in the early 19th century. The creation of differential and integral calculus initiated a period of rapid development in mathematics and in related applied disciplines.

Differential calculus is usually understood to mean classical differential calculus, which deals with real-valued functions of one or more real variables, but its modern definition may also include differential calculus in abstract spaces. Differential calculus is based on the concepts of real number; function; limit and continuity — highly important mathematical concepts, which were formulated and assigned their modern content during the development of mathematical analysis and during studies of its foundations. The central concepts of differential calculus — the derivative and the differential — and the apparatus developed in this connection furnish tools for the study of functions which locally look like linear functions or polynomials, and it is in fact such functions which are of interest, more than other functions, in applications.

Second Order Derivative

For understanding the second-order derivative, let us step back a bit and understand what a first derivative is. The first derivative $\frac{dy}{dx}$ represents the rate of the change in y with respect to x. Considering an example, if the distance covered by a car in 10 seconds is 60 meters, then the speed is the first order derivative of the distance travelled with respect to time. Hence, the speed in this case is given as $\frac{60}{10}$ m / s. We know that speed also varies and does not remain constant forever. Thus, to measure this rate of change in speed, one can use the second derivative.

The second-order derivative is nothing but the derivative of the first derivative of the given function. So, the variation in speed of the car can be found out by finding out the second derivative, i.e. the rate of change of speed with respect to time (the second derivative of distance travelled with respect to the time).

Graphically the first derivative represents the slope of the function at a point, and the second derivative describes how the slope changes over the independent variable in the graph. For a function having a variable slope, the second derivative explains the curvature of the given graph.

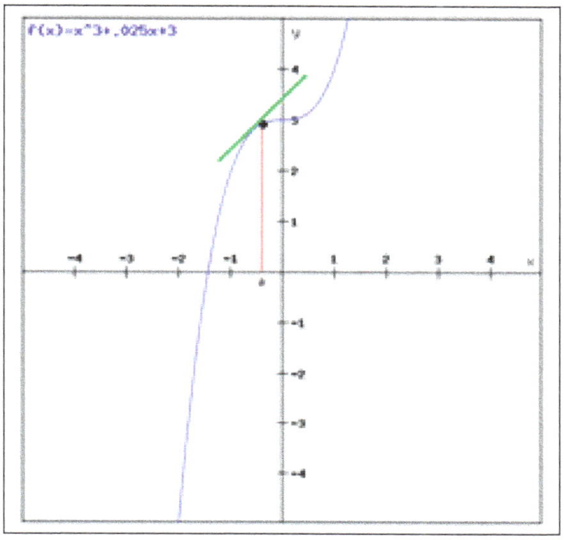

In this graph, the blue line indicates the slope, i.e. the first derivative of the given function. And the second derivative is used to define the nature of the given function. For example, we use the second derivative test to determine the maximum, minimum or the point of inflexion.

Mathematically, if $y = f(x)$,

Then $\frac{}{dx} \quad f'(x)$

Now if f'(x) is differentiable, then differentiating $\frac{dy}{dx}$ again w.r.t. x we get 2nd order derivative, i.e.

$$\frac{d}{dx}\left(\frac{dy}{dx}\right) = \frac{d^2y}{dx^2} = f''(x).$$

Similarly, higher order derivatives can also be defined in the same way like $\frac{d^3y}{dx^3}$ represents a third order derivative, $\frac{d^4y}{dx^4}$ represents a fourth order derivative and so on.

Usually, the second derivative of a given function corresponds to the curvature or concavity of the graph. If the second-order derivative value is positive, then the graph of a function is upwardly concave. If the second-order derivative value is negative, then the graph of a function is downwardly open. The second derivative of a function determines the local maximum or minimum, inflexion point values. These can be identified with the help of below conditions:

- If f''(x) < o, then the function f(x) has a local maximum at x.

- If f''(x) > 0, then the function f(x) has a local minimum at x.
- If f''(x) = 0, then it is not possible to conclude anything about the point x, a possible inflexion point.

Second Order Derivative Examples

Let us see an example to get acquainted with second-order derivatives.

Example: Find $\frac{d^2y}{dx^2}$ if $y = e^{(x^3)} - 3x^4$.

Solution: Given that $y = e^{(x^3)} - 3x^4$, then differentiating this equation w.r.t. x we get,

$$\frac{dy}{dx} = e^{(x^3)} \times 3x^2 - 12x^3$$

Now to find the 2nd order derivative of the given function, we differentiate the first derivative again w.r.t. x,

$$\frac{d^2y}{dx^2} = e^{(x^3)} \times 3x^2 \times 3x^2 + e^{(x^3)} \times 6x - 36x^2$$

$$\frac{d^2y}{dx^2} = xe^{(x^3)} \times \left(9x^3 + 6\right) - 36x^2$$

This is the required solution.

Example: Find $\frac{d^2y}{dx^2}$ if y = 4 $\sin^{-1}(x^2)$.

Solution: Given that y = 4 $\sin^{-1}(x^2)$, then differentiating this equation w.r.t. x we get,

$$\frac{dy}{dx} = \frac{4}{\sqrt{1-x^4}} \times 2x$$

Now for finding the next higher order derivative of the given function, we need to differentiate the first derivative again w.r.t. x,

$$\frac{d^2y}{dx^2} = 2x \times \frac{d}{dx}\left(\frac{4}{\sqrt{1-x^4}}\right) + \frac{4}{\sqrt{1-x^4}}\frac{d(2x)}{dx} \quad \left(\text{using } \frac{d(uv)}{dx} = u\frac{dv}{dx} + v\frac{du}{dx}\right)$$

$$\Rightarrow \frac{d^2y}{dx^2} = \frac{-8\left(x^4+1\right)}{\left(x^4-1\right)\sqrt{1-x^4}}$$

Derivation of nth Differential Coefficient of Some Standard Functions

1. n^{th} differential coefficient of x^m, where m is a positive integer ≥ n.

Let $y = x^m$.

Differentiating both sides w. r. t. x successively, we get,

$y_1 = m\,x^{m-1}$

$y_2 = m(m-1)\,x^{m-2}$

$y_3 = m(m-1)(m-2)\,x^{m-3}$

....

....

Now we suppose that,

$$y_n = m(m-1)(m-2)(m-3).....(m-(n-1))x^{m-n}$$

Differentiating both sides of above equation w. r. t. x, we get,

$$y_{n+1} = m(m-1)(m-2)(m-3).....(m-(n-1))(m-n)x^{m-(n+1)}$$

Clearly, y_{n+1} is of the same form as y_n, given by equation $y_n = m(m-1)(m-2)(m-3).....(m-(n-1))x^{m-n}$. This ⇒ If the previous equation is true for a particular value of n, then previous equation is true for next higher value of n also.

Equation $y_n = m(m-1)(m-2)(m-3).....(m-(n-1))x^{m-n}$ is true for n = 1, 2, 3. Since previous this equation is true for n = 3, it is true for n = 4 also. Similarly, since previous equation is true for n = 4, it is true for n = 5 also and so on. Therefore, by mathematical induction, previous equation is true for every positive integer n.

From previous equation we can write y_n as follows:

$$y_n = \frac{m(m-1)(m-2)(m-3).....(m-(n-1))(m-n)(m-(n+1))....3.2.1}{(m-n)(m-(n+1))....3.2.1}x^{m-n}$$

i.e., $\boxed{y_n = \frac{m!}{(m-n)!}x^{m-n}}$

Corollary: If $y = x^n$, then $y_n = n!$

2. n^{th} differential coefficient of m $(ax + b)^m$, where m is a positive integer ≥ n.

Let $y = (ax + b)^m$

Differentiating both sides w. r. t. x successively, we get,

$y_1 = ma(ax+b)^{m-1}$

$y_2 = m(m-1)a^2(ax+b)^{m-2}$

$y_3 = m(m-1)(m-2)a^3(ax+b)^{m-3}$

....

....

Now we suppose that,

$$y_n = m(m-1)(m-2)(m-2).....(m-(n-1))a^n(ax+b)^{m-n}$$

Differentiating both sides of above equation w. r. t. x, we get,

$$y_{n+1} = m(m-1)(m-2)(m-3).....(m-(n-1))(m-n)a^{n+1}(ax+b)^{m-(n+1)}$$

Clearly, y_{n+1} is of the same form as y_n, given by equation $y_n = m(m-1)(m-2)(m-2).....(m-(n-1))a^n(ax+b)^{m-n}$. This ⇒ If previous equation is true for a particular value of n, then previous equation is true for next higher value of n also.

Previous equation is true for n = 1, 2, 3. Since previous equation is true for n = 3, it is true for n = 4 also. Similarly, since previous equation is true for n = 4, it is true for n = 5 also and so on. Therefore, by mathematical induction, previous equation is true for every positive integer n.

From previous equation, we can write y_n as follows:

$$y_n = \frac{m(m-1)(m-2)(m-2).....(m-(n-1))(m-n)(m-n-1)....3.2.1}{(m-n)(m-n-1)....3.2.1}a^n(ax+b)^{m-n}$$

i.e., $\boxed{y_n = \frac{m!}{(m-n)!}a^n(ax+b)^{m-n}}$

3. n^{th} differential coefficient of $\frac{1}{ax+b}$.

Let $y = \frac{1}{ax+b} = (ax+b)^{-1}$.

Differentiating y w. r. t. x successively, we get,

$$y_1 = (-1)a(ax+b)^{-2}$$
$$y_2 = (-1)(-2)a^2(ax+b)^{-3}$$
$$y_3 = (-1)(-2)(-3)a^3(ax+b)^{-4}$$

....

....

Now we suppose that,

$$y_n = (-1)(-2)(-3).... (-n)a^n(ax+b)^{-(n+1)}$$

Differentiating both sides of above equation w. r. t. x, we get,

$$y_{n+1} = (-1)(-2)(-3).... (-n)(-(n+1))a^{n+1}(ax+b)^{-(n+2)}$$

Clearly, y_{n+1} is of the same form as y_n, given by equation $y_n = (-1)(-2)(-3)....(-n)a^n(ax+b)^{-(n+1)}$.
This ⇒ If previous equation is true for a particular value of n, then previous equation is true for next higher value of n also.

Previous equation is true for n = 1, 2, 3. Since previous equation is true for n = 3, it is true for n = 4 also. Similarly, since previous equation is true for n = 4, it is true for n = 5 also and so on. Therefore, by mathematical induction, previous equation is true for every positive integer n.

From previous equation, we can write y n as follows:

$$y_n = (-1)^n n! a^n (ax+b)^{-(n+1)}$$

i.e., $\boxed{y_n = \dfrac{(-1)^n a^n n!}{(ax+b)^{(n+1)}}}$

4. nth differential coefficient of $\dfrac{1}{(ax+b)^2}$

Let $y = \dfrac{1}{(ax+b)^2} = (ax+b)^{-2}$

Differentiating both sides w. r. t. x successively, we get,

$$y_1 = (-2)a(ax+b)^{-3}$$
$$y_2 = (-2)(-3)a^2(ax+b)^{-4}$$
$$y_3 = (-2)(-3)(-4)a^3(ax+b)^{-5}$$

....

....

Now we suppose that,

$$y_n = (-2)(-3)(-4)....(-(n+1))a^n(ax+b)^{-(n+2)}$$

Differentiating both sides of above equation w. r. t. x, we get,

$$y_{n+1} = (-2)(-3)(-4)....(-(n+1))(-(n+2))a^{n+1}(ax+b)^{-(n+3)}$$

Clearly, y_{n+1} is of the same form as y_n, given by equation $y_n = (-2)(-3)(-4)....(-(n+1))a^n(ax+b)^{-(n+2)}$.
This ⇒ If previous equation is true for a particular value of n, then previous equation is true for next higher value of n also.

Previous equation is true for n = 1, 2, 3. Since previous equation is true for n = 3, it is true for n = 4 also. Similarly, since previous equation is true for n = 4, it is true for n = 5 also and so on. Therefore, by mathematical induction, previous equation is true for every positive integer n. From previous equation, we can write y n as follows:

$$y_n = (-1)^n (n+1)! a^n (ax+b)^{-(n+2)}$$

i.e., $\boxed{y_n = \dfrac{(-1)^n a^n (n+1)!}{(ax+b)^{(n+2)}}}$

5. n^{th} differential coefficient of log (ax + b).

Let y = log (ax + b).

Differentiating both sides w. r. t. x successively, we get,

$$y_1 = \frac{a}{ax+b} = a(ax+b)^{-1}$$
$$y_2 = (-1)a^2(ax+b)^{-2}$$
$$y_3 = (-1)(-2)a^3(ax+b)^{-3}$$

....

....

Now we suppose that,

$$y_n = (-1)(-2)(-3)....(-(n-1))a^n(ax+b)^{-n}$$

Differentiating both sides of above equation w. r. t. x, we get,

$$y_{n+1} = (-1)(-2)(-3)....(-(n-1))(-n)a^{n+1}(ax+b)^{-(n+1)}$$

Clearly, y_{n+1} is of the same form as y_n, given by equation $y_n = (-1)(-2)(-3)....(-(n-1))a^n(ax+b)^{-n}$.

This ⇒ If previous equation is true for a particular value of n, then previous equation is true for next higher value of n also.

Previous equation is true for n = 1, 2, 3. Since previous equation is true for n = 3, it is true for n = 4 also. Similarly, since previous equation is true for n = 4, it is true for n = 5 also and so on. Therefore, by mathematical induction, previous equation is true for every positive integer n.

From previous equation, we can write y_n as follows:

$$y_n = (-1)^{n-1}(n-1)!a^n(ax+b)^{-n}$$

i.e., $\boxed{y_n = \dfrac{(-1)^{n-1} a^n (n-1)!}{(ax+b)^n}}$

6. n^{th} differential coefficient of e^{mx}.

Let $y = e^{mx}$.

Differentiating both sides w. r. t. x, successively, we get,

$y_1 = me^{mx}$
$y_2 = m^2e^{mx}$
$y_3 = m^3e^{mx}$
....
....

Now we suppose that,

$$y_n = m^n e^{mx}$$

Differentiating both sides of above equation w. r. t. x we get,

$$y_{n+1} = m^n\left(me^{mx}\right) \text{ i.e, } y_{n+1} = m^{n+1}e^{mx}$$

Clearly, y_{n+1} is of the same form as y_n, given by equation $y_n = m^n e^{mx}$. This ⇒ If previous equation is true for a particular value of n, then previous equation is true for next higher value of n also.

Previous equation is true for n = 1, 2, 3. Since previous equation is true for n = 3, it is true for n = 4 also. Similarly, since previous equation is true for n = 4, it is true for n = 5 also and so on. Therefore, by mathematical induction, previous equation is true for every positive integer n,

i.e., $\boxed{y_n = m^n e^{mx}}$

7. n^{th} differential coefficient of a^{mx}.

Let $y = a^{mx}$.

Then $y = e^{mx\log a}$ $\left(\because a^{mx} = e^{mx\log a}\right)$.

Differentiating both sides w. r. t. x successively, we get,

$y_1 = m\log a\left(e^{mx\log a}\right)$
$y_2 = m\log a\left\{m\log a\left(e^{mx\log a}\right)\right\} = (m\log a)^2\left(e^{mx\log a}\right)$
$y_3 = (m\log a)^2\left\{m\log a\left(e^{mx\log a}\right)\right\} == (m\log a)^3\left(e^{mx\log a}\right)$
....
....

Now we suppose that,

$$y_n = (m\log a)^n\left(e^{mx\log a}\right).$$

Differentiating both sides of above equation w. r. t. x, we get,

$$y_{n+1} = (m\log a)^{\bar{n}}\left\{m\log a\left(e^{mx\log a}\right)\right\} == (m\log a)^{n+1}\left(e^{mx\log a}\right)$$

Clearly, y_{n+1} is of the same form as y_n, given by equation $y_n = (m\log a)^n (e^{mx\log a})$. This ⇒ If previous equation is true for a particular value of n, then previous equation is true for next higher value of n also.

Previous equation is true for n = 1, 2, 3. Since previous equation is true for n = 3, it is true for n = 4 also. Similarly, since previous equation is true for n = 4, it is true for n = 5 also and so on. Therefore, by mathematical induction, previous equation is true for every positive integer n,

$$y_n = (m\log a)^n (e^{mx\log a})$$

i.e., $y_n = (m\log a)^n a^{mx} \quad (\because a^{mx} = e^{mx\log a})$

i.e., $\boxed{y_n = m^n (\log a)^n a^{mx}}$

8. n^{th} differential coefficient of sin (ax + b).

Let y = sin(ax + b).

Differentiating both sides w. r. t. x successively, we get,

$$y_1 = a\cos(ax+b) = a\sin\left(ax+b+\frac{\pi}{2}\right)$$

$$y_2 = a^2\cos\left(ax+b+\frac{\pi}{2}\right) = a^2\sin\left(ax+b+\frac{\pi}{2}+\frac{\pi}{2}\right) = a^2\sin\left(ax+b+2\frac{\pi}{2}\right)$$

$$y_3 = a^3\cos\left(ax+b+2\frac{\pi}{2}\right) = a^3\sin\left(ax+b+2\frac{\pi}{2}+\frac{\pi}{2}\right) = a^3\sin\left(ax+b+3\frac{\pi}{2}\right)$$

....

....

Now we suppose that,

$$y_n = a^n\sin\left(ax+b+n\frac{\pi}{2}\right)$$

Differentiating both sides of above equation w. r. t. x, we get,

$$y_{n+1} = a^{n+1}\cos\left(ax+b+n\frac{\pi}{2}\right) = a^{n+1}\sin\left(ax+b+n\frac{\pi}{2}+\frac{\pi}{2}\right)$$

i.e., $y_{n+1} = a^{n+1}\sin\left(ax+b+(n+1)\frac{\pi}{2}\right)$

Clearly, y_{n+1} is of the same form as y_n, given by equation $y_n = a^n\sin\left(ax+b+n\frac{\pi}{2}\right)$. This ⇒ If previous equation is true for a particular value of n, then previous equation is true for next higher value of n also.

Previous equation is true for n = 1, 2, 3. Since previous equation is true for n = 3, it is true for n =

4 also. Similarly, since previous equation is true for n = 4, it is true for n = 5 also and so on. Therefore, by mathematical induction, previous equation is true for every positive integer n,

i.e., $\boxed{y_n = a^n \sin\left(ax+b+\frac{n\pi}{2}\right)}$

9. n^{th} differential coefficient of cos(ax + b).

Let y = cos(ax + b).

Differentiating both sides w. r. t. x successively, we get,

$$y_1 = -a\sin(ax+b) = a\cos\left(ax+b+\frac{\pi}{2}\right)$$

$$y_2 = -a^2 \sin\left(ax+b+\frac{\pi}{2}\right) = a^2\cos\left(ax+b+\frac{\pi}{2}+\frac{\pi}{2}\right) = a^2\cos\left(ax+b+2\frac{\pi}{2}\right)$$

$$y_3 = -a^3 \sin\left(ax+b+2\frac{\pi}{2}\right) = a^3\cos\left(ax+b+2\frac{\pi}{2}+\frac{\pi}{2}\right) = a^3\cos\left(ax+b+3\frac{\pi}{2}\right)$$

………………………………

………………………………

Now we suppose that,

$$y_n = a^n \cos\left(ax+b+n\frac{\pi}{2}\right)$$

Differentiating both sides of above equation w. r. t. x, we get,

$$y_{n+1} = -a^{n+1}\sin\left(ax+b+n\frac{\pi}{2}\right) = a^{n+1}\cos\left(ax+b+n\frac{\pi}{2}+\frac{\pi}{2}\right)$$

i.e., $y_{n+1} = a^{n+1}\cos\left(ax+b+(n+1)\frac{\pi}{2}\right)$

Clearly, y_{n+1} is of the same form as y_n, given by equation $y_n = a^n\cos\left(ax+b+n\frac{\pi}{2}\right)$. This ⇒ If previous equation is true for a particular value of n, then previous equation is true for next higher value of n also.

Previous equation is true for n = 1, 2, 3. Since previous equation is true for n = 3, it is true for n = 4 also. Similarly, since previous equation is true for n = 4, it is true for n = 5 also and so on. Therefore, by mathematical induction, previous equation is true for every positive integer n,

i.e., $\boxed{y_n = a^n \cos\left(ax+b+\frac{n\pi}{2}\right)}$

10. n^{th} differential coefficient of $e^{ax}\sin(bx + c)$.

Let $y = e^{ax}\sin(bx + c)$.

Differentiating both sides w. r. t. x, we get,

$$y_1 = a e^{ax} \sin(bx+c) + b e^{ax} \cos(bx+c)$$

Put a = r cosθ and b = rsinθ so that $r = \sqrt{a^2+b^2}$ and $\theta = \tan^{-1}\frac{b}{a}$

Then we have,

$$\begin{aligned} y_1 &= r\cos\theta e^{ax}\sin(bx+c) + r\sin\theta e^{ax}\cos(bx+c) \\ &= r e^{ax}\left(\sin(bx+c)\cos\theta + \cos(bx+c)\sin\theta\right) \\ &= r e^{ax}\sin(bx+c+\theta) \end{aligned}$$

Similarly, differentiating y w. r. t. x twice, thrice, ..., we get,

$$y_2 = r^2 e^{ax}\sin(bx+c+2\theta)$$
$$y_3 = r^3 e^{ax}\sin(bx+c+3\theta)$$

....

....

Now we suppose that,

$$y_n = r^n e^{ax}\sin(bx+c+n\theta)$$

Differentiating both sides of above equation w. r. t. x, we get,

$$\begin{aligned} y_{n+1} &= r^n\left[a e^{ax}\sin(bx+c+n\theta) + b e^{ax}\cos(bx+c+n\theta)\right] \\ &= r^n e^{ax}\left[a\sin(bx+c+n\theta) + b\cos(bx+c+n\theta)\right] \\ &= r^n e^{ax}\left[r\cos\theta\ \sin(bx+c+n\theta) + r\sin\theta\cos(bx+c+n\theta)\right] \\ &\qquad (\text{using } a = r\cos\theta,\ b = \sin\theta) \\ &= r^{n+1} e^{ax}\left[\sin(bx+c+n\theta) + \cos\theta + \cos(bx+c+n\theta)\sin\theta\right] \\ &= r^{n+1} e^{ax}\sin(bx+c+n\theta+\theta) \\ &= r^{n+1} e^{ax}\sin(bx+c+(n+1)\theta) \end{aligned}$$

Clearly, y_{n+1} is of the same form as y_n, given by equation $y_n = r^n e^{ax}\sin(bx+c+n\theta)$. This ⇒ If previous equation is true for a particular value of n, then previous equation is true for next higher value of n also.

Previous equation is true for n = 1, 2, 3. Since previous equation is true for n = 3, it is true for n = 4 also. Similarly, since previous equation is true for n = 4, it is true for n = 5 also and so on. Therefore, by mathematical induction, equation previous is true for every positive integer n,

i.e., $\boxed{y_n = r^n e^{ax}\sin(bx+c+n\theta),\ \text{where } r = \sqrt{a^2+b^2} \text{ and } \theta = \tan^{-1}\frac{b}{a}}$

11. n^{th} differential coefficient of $e^{ax}\cos(bx+c)$.

Let $y = e^{ax}\cos(bx+c)$.

Differentiating both sides w. r. t. x, we get,

$$y_1 = ae^{ax}\cos(bx+c) - be^{ax}\sin(bx+c)$$

Put a = r cosθ and b = r sinθ so that $r = \sqrt{a^2+b^2}$ and $\theta = \tan^{-1}\dfrac{b}{a}$.

Then we have,

$$\begin{aligned} y_1 &= r\cos\theta e^{ax}\cos(bx+c) - r\sin\theta e^{ax}\sin(bx+c) \\ &= re^{ax}\left(\cos(bx+c)\cos\theta - \sin(bx+c)\sin\theta\right) \\ &= re^{ax}\cos(bx+c+\theta) \end{aligned}$$

Similarly, differentiating y w. r. t. x twice, thrice, ..., we get,

$$y_2 = r^2 e^{ax}\cos(bx+c+2\theta)$$
$$y_3 = r^3 e^{ax}\cos(bx+c+3\theta)$$

....

....

Now we suppose that,

$$y_n = r^n e^{ax}\cos(bx+c+n\theta)$$

Differentiating both sides of above equation w. r. t. x, we get,

$$\begin{aligned} y_{n+1} &= r^n\left[ae^{ax}\cos(bx+c+n\theta) - be^{ax}\sin(bx+c+n\theta)\right] \\ &= r^n e^{ax}\left[a\cos(bx+c+n\theta) - b\sin(bx+c+n\theta)\right] \\ &= r^n e^{ax}\left[r\cos\theta\cos(bx+c+n\theta) - r\sin\theta\sin(bx+c+n\theta)\right] \\ &\qquad (\text{using } a = r\cos\theta,\ b = \sin\theta) \\ &= r^{n+1}e^{ax}\left[\cos(bx+c+n\theta) + \cos\theta - \sin(bx+c+n\theta)\sin\theta\right] \\ &= r^{n+1}e^{ax}\cos(bx+c+n\theta+\theta) \\ &= r^{n+1}e^{ax}\cos(bx+c+(n+1)\theta) \end{aligned}$$

Clearly, y_{n+1} is of the same form as y_n, given by equation $y_n = r^n e^{ax}\cos(bx+c+n\theta)$. This ⇒ If previous equation is true for a particular value of n, then previous equation is true for next higher value of n also.

Previous equation is true for n = 1, 2, 3. Since previous equation is true for n = 3, it is true for n = 4 also. Similarly, since previous equation is true for n = 4, it is true for n = 5 also and so on. Therefore, by mathematical induction, previous equation is true for every positive integer n, i.e., $\boxed{y_n = r^n e^{ax} \cos(bx + c + n\theta), \text{ where } r = \sqrt{a^2 + b^2} \text{ and } \theta = \tan^{-1}\frac{b}{a}}$

Example: Find n^{th} differential coefficient of $\frac{1}{(1+x)^2}$.

Solution: Let $y = \frac{1}{(1+x)^2}$.

Differentiating both sides w. r. t. x successively, we get,

$$y_1 = (-2)(1+x)^{-3}$$
$$y_2 = (-2)(-3)(1+x)^{-4}$$
$$y_3 = (-2)(-3)(-4)(1+x)^{-5}$$

....

....

Proceeding in this way, we get,

$$y_n = (-2)(-3)(-4)....(-(n+1))(1+x)^{-(n+2)}$$

i.e., $y_n = \frac{(-1)^n (n+1)!}{(1+x)^{(n+2)}}$.

Example: Find y_n if $y = \sin^2 x \cos^2 x$.

Solution: We have,

$$y = \sin^2 \cos^2 x = \frac{1}{4}(2 \sin x \cos x)^2 = \frac{1}{4}(\sin 2x)^2 = \frac{1}{4}\sin^2 2x = \frac{1}{4}\left(\frac{1-\cos 4x}{2}\right)$$

i.e., $y = \frac{1}{8} - \frac{1}{8}\cos 4x$

This $\Rightarrow y_n = 0 - \frac{1}{8}4^n \cos\left(4x + \frac{n\pi}{2}\right)$

(using the formula: $y_n = a^n \cos\left(ax + b + \frac{n\pi}{2}\right)$ if $y = \cos(ax + b)$)

i.e., $y_n == -2^{n-2} \cos\left(4x + \frac{n\pi}{2}\right)$.

Example: Find y_n if $y = e^x \sin^2 x$.

Solution: We have,

$$y = e^x \sin^2 x = e^x\left(\frac{1-\cos 2x}{2}\right)$$

i.e., $y = \frac{1}{2}e^x - \frac{1}{2}e^x \cos 2x$

This $\Rightarrow y_n = \frac{1}{2}e^x - \frac{1}{2}\left(1^2+2^2\right)^{\frac{n}{2}} e^x \cos\left(2x + n\tan^{-1} 2\right)$

$$\left(\begin{array}{lll}\text{using the formulae:} & y_n = m^n e^{mx} \text{ if } y = e^{mx} \\ \text{and} & y_n = \left(a^2+b^2\right)^{\frac{n}{2}} e^{ax} \cos\left(bx + c + n\tan^{-1}\frac{b}{a}\right) \text{ if } y = e^{ax}\cos(bx+c)\end{array}\right)$$

i.e., $y_n = \frac{1}{2}e^x\left[1-(5)^{\frac{n}{2}}\cos\left(2x+n\tan^{-1}2\right)\right].$

Example: If y = sin mx + cos mx, prove that $y_n = m^n\left[1+(-1)^n \sin 2mx\right]^{\frac{1}{2}}$.

Solution: We have,

$$y = \sin mx + \cos mx$$

This $\Rightarrow y_n = m^n \sin\left(mx + \frac{n\pi}{2}\right) + m^n \cos\left(mx + \frac{n\pi}{2}\right)$

$$\left(\begin{array}{lll}\text{using the formulae:} & y_n = a^n \sin\left(ax+b+\frac{n\pi}{2}\right) \text{if } y = \sin(ax+b) \\ \text{and} & y_n = a^n \cos\left(ax+b+\frac{n\pi}{2}\right) \text{ if } y = \cos(ax+b)\end{array}\right)$$

i.e., $y_n = m^n\left\{\sin\left(mx+\frac{n\pi}{2}\right)+\cos\left(mx+\frac{n\pi}{2}\right)\right\}.$

This $\Rightarrow y_n = m^n\left[\left\{\sin\left(mx+\frac{n\pi}{2}\right)+\cos\left(mx+\frac{n\pi}{2}\right)\right\}^2\right]^{\frac{1}{2}}$

$$= m^n\left[\sin^2\left(mx+\frac{n\pi}{2}\right)+\cos^2\left(mx+\frac{n\pi}{2}\right)+2\sin\left(mx+\frac{n\pi}{2}\right)\cos\left(mx+\frac{n\pi}{2}\right)\right]^{\frac{1}{2}}$$

$$= m^n\left[1+2\sin\left(mx+\frac{n\pi}{2}\right)\cos\left(mx+\frac{n\pi}{2}\right)\right]^{\frac{1}{2}}$$

$$= m^n\left[1+\sin 2\left(mx+\frac{n\pi}{2}\right)\right]^{\frac{1}{2}} \qquad \text{(using the formulae } 2\sin\theta\cos\theta = \sin 2\theta\text{)}$$

$$= m^n\left[1+\sin(2mx+n\pi)\right]^{\frac{1}{2}}$$

$$= m^n\left[1+\sin 2mx \cos n\pi + \cos 2mx \sin n\pi\right]^{\frac{1}{2}}$$

$$= m^n\left[1+(-1)^n \sin 2mx\right]^{\frac{1}{2}} \qquad \left(\text{since } \cos n\pi = (-1)^n \text{ and } \sin n\pi = 0\right)$$

Method of Finding n^{th} Derivative of an Algebraic Rational Function

In order to find the nth derivative of an algebraic rational function, it is first required to check whether the denominator of the given algebraic rational function is resolvable into real linear factors or not. If the denominator of the given algebraic rational function is resolvable into real linear factors, then the given algebraic rational function is resolved into partial fractions.

In this case, the denominator of each partial fraction thus obtained consists of real linear factors, repeated or non-repeated, therefore the nth derivative of the given algebraic rational function can be found out by directly using the formulae derived earlier for nth derivative of some standard functions. If the denominator of the given algebraic rational function is not resolvable into real linear factors, then its nth derivative is found out by the application of De Moivre's theorem.

Example: Find y_n if $y=\frac{x^2}{(x-1)^2(x+2)}$.

Solution: We have,

$$y=\frac{x^2}{(x-1)^2(x+2)}$$

Let us resolve $\frac{x^2}{(x-1)^2(x+2)}$ into partial fractions.

Let $\frac{x^2}{(x-1)^2(x+2)}=\frac{A}{x-1}+\frac{B}{(x-1)^2}+\frac{C}{x+2}$.

This $\Rightarrow \frac{x^2}{(x-1)^2(x+2)}=\frac{A(x-1)(x+2)+B(x+2)+C(x-1)^2}{(x-1)^2(x+2)}$

$$\Rightarrow x^2 = A(x-1)(x+2)+B(x+2)+C(x-1)^2$$

Putting x = 1 on both sides of above equation, we get B = 1/3.

Putting x = - 2 on both sides of above equation, we get C = 4/9.

Equating the coefficients of x^2 from both sides of above equation, we get,

$$1=A+C, \text{ which} \Rightarrow A=1-C=1-\frac{4}{9}=\frac{5}{9}.$$

Substituting the values of A, B and C in equation $\frac{x^2}{(x-1)^2(x+2)}=\frac{A}{x-1}+\frac{B}{(x-1)^2}+\frac{C}{x+2}$ and using equation $y=\frac{x^2}{(x-1)^2(x+2)}$, we get,

$$y=\frac{5}{9}\left(\frac{1}{x-1}\right)+\frac{1}{3(x-1)^2}+\frac{4}{9}\left(\frac{1}{x+2}\right)$$

This $\Rightarrow y_n=\frac{5}{9}\left[\frac{(-1)^n n!}{(x-1)^{n+1}}\right]+\frac{1}{3}\left[\frac{(-1)^n (n+1)!}{(x-1)^{n+2}}\right]+\frac{4}{9}\left[\frac{(-1)^n n!}{(x+2)^{n+1}}\right]$

$$\left(\begin{array}{ll}\text{using the formulae: } & y_n = \dfrac{(-1)^n a^n n!}{(ax+b)^{n+1}} \text{ if } y = \dfrac{1}{ax+b} \\ \text{and} & y_n = \dfrac{(-1)^n a^n (n+1)!}{(ax+b)^{n+2}} \text{ if } y = \dfrac{1}{(ax+b)^2}\end{array}\right)$$

Example: Find y_n if $y = \dfrac{1}{x^2 + a^2}$.

Solution: We have,

$$y = \frac{1}{x^2 + a^2} = \frac{1}{(x+ia)(x-ia)} = \frac{1}{2ia}\left[\frac{(x+ia)-(x-ia)}{(x+ia)(x-ia)}\right]$$

i.e., $y = \dfrac{1}{2ia}\left[\dfrac{1}{(x-ia)} - \dfrac{1}{(x+ia)}\right]$

This $\Rightarrow y_n = \dfrac{1}{2ia}\left[\dfrac{(-1)^n n!}{(x-ia)^{n+1}} - \dfrac{(-1)^n n!}{(x+ia)^{n+1}}\right]\left(\text{using the formula: } y_n = \dfrac{(-1)^n a^n n!}{(ax+b)^{n+1}} \text{ if } y = \dfrac{1}{ax+b}\right)$

i.e., $y_n = \dfrac{(-1)^n n!}{2ia}\left[\dfrac{1}{(x-ia)^{n+1}} - \dfrac{1}{(x+ia)^{n+1}}\right]$

Now put x = r cos θ and a = r sin θ so that $r = \sqrt{x^2 + a^2}$ and $\theta = \tan^{-1}\dfrac{a}{x}$.

Then we have,

$$y_n = \frac{(-1)^n n!}{2ia}\left[\frac{1}{(r\cos\theta - ir\sin\theta)^{n+1}} - \frac{1}{(r\cos\theta + ir\sin\theta)^{n+1}}\right]$$

This $\Rightarrow y_n = \dfrac{(-1)^n n!}{2iar^{n+1}}\left[\dfrac{1}{(\cos\theta - i\sin\theta)^{n+1}} - \dfrac{1}{(\cos\theta + i\sin\theta)^{n+1}}\right]$

$$= \frac{(-1)^n n!}{2iar^{n+1}}\left[\frac{1}{\left(e^{-i\theta}\right)^{n+1}} - \frac{1}{\left(e^{i\theta}\right)^{n+1}}\right] \quad \left(\text{using the formula: } e^{i\theta} = \cos\theta + i\sin\theta\right)$$

$$= \frac{(-1)^n n!}{2iar^{n+1}}\left[\frac{1}{e^{-i(n+1)\theta}} - \frac{1}{e^{i(n+1)\theta}}\right]$$

$$= \frac{(-1)^n n!}{2iar^{n+1}}\left[e^{i(n+1)\theta} - e^{-i(n+1)\theta}\right]$$

$$= \frac{(-1)^n n!}{2iar^{n+1}}\left[\{\cos(n+1)\theta + i\sin(n+1)\theta\} - \{\cos(n+1)\theta - i\sin(n+1)\theta\}\right]$$

$$= \frac{(-1)^n n!}{2ia\,r^{n+1}}\left[2i\sin(n+1)\theta\right]$$

$$= \frac{(-1)^n n!}{a\left(\dfrac{a}{\sin\theta}\right)^{n+1}}\sin(n+1)\theta \qquad \left(\text{since } a = r\sin\theta, \text{ i.e., } r = \frac{a}{\sin\theta}\right)$$

i.e., $y_n = \dfrac{(-1)^n n!}{a^{n+2}}\sin\theta^{n+1}\theta\sin(n+1)\theta.$

Example: Find y_n if $y = \tan^{-1} x$.

Solution: We have $y = \tan^{-1} x$.

Differentiating both sides w. r. t. x, we get,

$$y_1 = \frac{1}{1+x^2}$$

This $\Rightarrow y_1 == \dfrac{1}{(x+i)(x-i)} = \dfrac{1}{2i}\left[\dfrac{(x+i)-(x-i)}{(x+i)(x-i)}\right]$

i.e., $y_1 = \dfrac{1}{2i}\left[\dfrac{1}{(x-i)} - \dfrac{1}{(x+i)}\right]$

Differentiating y_1 w. r. t. x (n −1) times, we get,

$$y_n = \frac{1}{2i}\left[\frac{d^{n-1}}{dx^{n-1}}\left(\frac{1}{(x-i)}\right) - \frac{d^{n-1}}{dx^{n-1}}\left(\frac{1}{(x+i)}\right)\right]$$

This $\Rightarrow y_n = \dfrac{1}{2i}\left[\dfrac{(-1)^{n-1}(n-1)!}{(x-i)^n} - \dfrac{(-1)^{n-1}(n-1)!}{(x+i)^n}\right]$

$$\left(\text{using the formula}: y_n = \frac{(-1)^n a^n n!}{(ax+b)^{n+1}} \text{ if } y = \frac{1}{ax+b}\right)$$

i.e., $y_n = \dfrac{(-1)^{n-1}(n-1)!}{2i}\left[\dfrac{1}{(x-i)^n} - \dfrac{1}{(x+i)^n}\right].$

Now put $x = r\cos\theta$ and $1 = r\sin\theta$ so that $r = \sqrt{x^2+1}$ and $\theta = \tan^{-1}\dfrac{1}{x}$.

Then we have,

$$y_n = \frac{(-1)^{n-1}(n-1)!}{2i}\left[\frac{1}{(r\cos\theta - ir\sin\theta)^n} - \frac{1}{(r\cos\theta + ir\sin\theta)^n}\right]$$

$$\text{This} \Rightarrow y_n = \frac{(-1)^{n-1}(n-1)!}{2ir^n}\left[\frac{1}{(\cos\theta - i\sin\theta)^n} - \frac{1}{(\cos\theta + i\sin\theta)^n}\right]$$

$$= \frac{(-1)^{n-1}(n-1)!}{2ir^n}\left[\frac{1}{\left(e^{-i\theta}\right)^n} - \frac{1}{\left(e^{i\theta}\right)^n}\right] \quad \left(\text{using } e^{i\theta} = \cos\theta + i\sin\theta\right)$$

$$= \frac{(-1)^{n-1}(n-1)!}{2ir^n}\left[\frac{1}{e^{-in\theta}} - \frac{1}{e^{in\theta}}\right]$$

$$= \frac{(-1)^{n-1}(n-1)!}{2ir^n}\left[e^{in\theta} - e^{-in\theta}\right]$$

$$= \frac{(-1)^{n-1}(n-1)!}{2ir^n}\left[\{\cos n\theta + i\sin n\theta\} - \{\cos n\theta - i\sin n\theta\}\right]$$

$$= \frac{(-1)^{n-1}(n-1)!}{2ir^n}\left[2i\sin n\theta\right]$$

$$= \frac{(-1)^{n-1}(n-1)!}{\left(\frac{1}{\sin\theta}\right)^n}\sin n\theta \quad \left(\text{since } 1 = r\sin\theta, \text{ i.e., } r = \frac{1}{\sin\theta}\right)$$

i.e., $y \; = (-1)^{n \; 1}(n-1)!\sin \;\; \theta \sin n\theta.$

Higher Order Derivative

Because the derivative of a function y = f(x) is itself a function y′ = f′(x), you can take the derivative of f′(x), which is generally referred to as the second derivative of f(x) and written f′′(x) or $f^2(x)$. This differentiation process can be continued to find the third, fourth, and successive derivatives of f(x), which are called higher order derivatives of f(x).

Notation for Higher-Order Derivatives

	Prime Notation	Ratio Notation
1st Derivative	$f'(x)$	$\frac{df}{dx}$
2nd Derivative	$f''(x)$	$\frac{d^2f}{dx^2}$
3rd Derivative	$f'''(x)$	$\frac{d^3f}{dx^3}$
4th Derivative	$f^{(4)}(x)$	$\frac{d^4f}{dx^4}$

5th Derivative	$f^{(5)}(x)$	$\frac{d^5f}{dx^5}$
⋮	⋮	⋮
n^{th} Derivative	$f^{(n)}(x)$	$\frac{d^nf}{dx^n}$

After the third derivative, the prime notation changes. This is because it would be too hard to read lots of tick marks.

Example: Suppose $f(x) = 5x^3 + 3x^2 - 7x + 4$. Find the second and third derivatives.

Step 1: Find the first derivative.

$$f'(x) = 15x^2 + 6x - 7$$

Step 2: Find the second derivative.

$$f''(x) = 30x + 6$$

Step 3: Find the third derivative.

$$f'''(x) = 30$$

Answer: The second derivative is $f''(x) = 30x + 6$. The third derivative is $f'''(x) = 30$.

Example: Suppose $f(x) = \frac{4x+3}{2x-5}$. Find the second derivative.

Step 1: Find the first derivative using the quotient rule. Simplify the result.

$$\begin{aligned} f'(x) &= \frac{(2x-5)\cdot 4 - (4x+3)\cdot 2}{(2x-5)^2} \\ &= \frac{8x-20-(8x+6)}{(2x-5)^2} \\ &= \frac{8x-20-8x-6}{(2x-5)^2} \\ &= \frac{-20-6}{(2x-5)^2} \\ &= \frac{-26}{(2x-5)^2} \end{aligned}$$

Step 2: Rewrite the first derivative using a negative exponent.

$$f'(x) = -26(2x-5)^{-2}$$

Step 3: Find the second derivative using the chain rule. Simplify.

$$f''(x) = -26(-2)(2x-5)^{-3} \cdot \frac{d}{dx}(2x-5)$$
$$= 52(2x-5)^{-3} \cdot 2$$
$$= 104(2x-5)^{-3}$$
$$= \frac{104}{(2x-5)^3}$$

Answer: $f''(x) = \frac{104}{(2x-5)^3}$.

Leibniz Formula

The Leibniz formula expresses the derivative on nth order of the product of two functions. Suppose that the functions u(x) and v(x) have the derivatives up to nth order. Consider the derivative of the product of these functions.

The first derivative is described by the well-known formula:

$$(uv)' = u'v + uv'.$$

Differentiating this expression again yields the second derivative:

$$(uv)'' = \left[(uv)'\right]' = (u'v + uv')' = (u'v)' + (uv')' = u''v + u'v' + u'v' + uv'' = u''v + 2u'v' + uv''.$$

Likewise, we can find the third derivative of the product uv:

$$(uv)''' = \left[(uv)''\right]' = (u''v + 2u'v' + uv'')' = (u''v)' + (2u'v')' + (uv'')'$$
$$= u'''v + u''v' + 2u''v' + 2u'v'' + u'v'' + uv''' = u'''v + 3u''v' + 3u'v'' + uv'''.$$

It is easy to see that these formulas are similar to the binomial expansion raised to the appropriate exponent. Assuming that the terms with zero exponent u^o and v^o correspond to the functions u and v themselves, we can write the general formula for the derivative of nth order of the product of functions uv as follows:

$$(uv)^{(n)} = \sum_{i=o}^{n} \binom{n}{i} u^{(n-i)} v^{(i)},$$

where $\binom{n}{i}$ denotes the number of i-combinations of n elements.

This formula is called the Leibniz formula and can be proved by induction.

Proof: Suppose that the functions u and v have the derivatives of (n+1)th order. Using the recurrence relation, we write the expression for the derivative of (n+1)th order in the following form:

$$y^{(n+1)} = \left[y^{(n)}\right]' = \left[(uv)^{(n)}\right]' = \left[\sum_{i=o}^{n} \binom{n}{i} u^{(n-i)} v^{(i)}\right]'.$$

After differentiation we obtain:

$$y^{(n+1)} = \sum_{i=0}^{n} \binom{n}{i} u^{(n-i+1)} v^{(i)} + \sum_{i=0}^{n} \binom{n}{i} u^{(n-i)} v^{(i+1)}.$$

Both sums in the right-hand side can be combined into a single sum. Indeed, take an intermediate index $1 \le m \le n$. The first term when $i = m$ is written as,

$$\binom{n}{m} u^{(n-m+1)} v^{(m)},$$

and the second term when $i = m - 1$ is as follows:

$$\binom{n}{m-1} u^{(n-(m-1))} v^{((m-1)+1)} = \binom{n}{m-1} u^{(n-m+1)} v^{(m)}.$$

The sum of these two terms is given by,

$$\binom{n}{m} u^{(n-m+1)} v^{(m)} + \binom{n}{m-1} u^{(n-m+1)} v^{(m)} = \left[\binom{n}{m} + \binom{n}{m-1}\right] \cdot u^{(n-m+1)} v^{(m)}.$$

It is known from combinatorics that,

$$\binom{n}{m} + \binom{n}{m-1} = \binom{n+1}{m}.$$

Therefore, the sum of these two terms can be written as,

$$\left[\binom{n}{m} + \binom{n}{m-1}\right] \cdot u^{(n-m+1)} v^{(m)} = \binom{n+1}{m} u^{(n+1-m)} v^{(m)}.$$

It is clear that when m changes from 1 to n this combination will cover all terms of both sums except the term for $i = 0$ in the first sum equal to,

$$\binom{n}{0} u^{(n-0+1)} v^{(0)} = u^{(n+1)} v^{(0)},$$

and the term for $i = n$ in the second sum equal to,

$$\binom{n}{n} u^{(n-n)} v^{(n+1)} = u^{(0)} v^{(n+1)}.$$

As a result, the derivative of (n+1)th order of the product of functions uv is represented in the form,

$$y^{(n+1)} = u^{(n+1)} v^{(0)} + \sum_{m=1}^{n} \binom{n+1}{m} u^{(n+1-m)} v^{(m)} + u^{(0)} v^{(n+1)}$$

$$= \sum_{m=0}^{n+1} \binom{n+1}{m} u^{(n+1-m)} v^{(m)}.$$

The expression for $y^{(n+1)}$ has a similar form as for the derivative y(n). Only now the upper limit of summation is equal to n+1 instead of n. Thus, the Leibniz formula is proved for an arbitrary natural number n.

Example: If $y = a\cos(\log x) + b\sin(\log x)$, show that,

$$x^2y_{n+2} + (2n+1)xy_{n+1} + n(n+1)y_n = 0$$

Solution: Here $y = a\cos(\log x) + b\sin(\log x)$,

$$\Rightarrow y_1 = \frac{-a}{x}\sin(\log x) + \frac{b}{x}\cos(\log x)$$
$$\Rightarrow xy_1 = -a\sin(\log x) + b\cos(\log x)$$

Differentiating both sides w.r.t. x, we get,

$$xy_2 + y_1 = -\frac{a}{x}\cos(\log x) + \frac{-b}{x}\sin(\log x)$$
$$\Rightarrow x^2y_2 + xy_1 = -\{a\cos(\log x) + b\sin(\log x)\}$$
$$= -y$$
$$\Rightarrow x^2y_2 + xy_1 + y = 0$$

Using Leibnitz's theorem, we get,

$$\left(y_{n+2}x^2 + n_{c_1}y_{n+1}2x + n_{c_2}y_n \cdot 2\right) + \left(y_{n+1}x + n_{c_1}y_n \cdot 1\right) + y_n = 0$$
$$\Rightarrow y_{n+2}x^2 + y_{n+1}2nx + n(n-1)y_n + y_{n+1}x + ny_n + y_n = 0$$
$$\Rightarrow x^2y_{n+2} + (2n+1)xy_{n+1} + (n^2+1)y_n = 0$$

Example: If $y = \sin(m\sin^{-1}x)$, show that, $(1-x^2)y_{n+2} = (2n+1)xy_{n+1} + (n^2-m^2)y_n$. Also find $y_n(0)$.

Solution: Here $y = \sin(m\sin^{-1}x)$,

$$\Rightarrow y_1 = \frac{m}{\sqrt{1-x^2}}\cos(m\sin^{-1}x)$$
$$\Rightarrow (1-x^2)y_1^2 = m^2\cos^2(m\sin^{-1}x)$$
$$\Rightarrow (1-x^2)y_1^2 = m^2\left[1-\sin^2(m\sin^{-1}x)\right]$$
$$\Rightarrow (1-x^2)y_1^2 = m^2(1-y^2)$$
$$\Rightarrow (1-x^2)y_1^2 + m^2y^2 = m^2$$

Differentiating w.r.t. x, we get,

$$(1-x^2)2y_1y_2 + y_1^2(-2x) + m^2 2yy_1 = 0$$
$$\Rightarrow (1-x^2)y_2 - xy_1 + m^2y = 0$$

Using Leibnitz's theorem, we get,

$$\left[y_{n+2}\left(1-x^2\right)+n_{c_1}y_{n+1}\left(-2x\right)+n_{c_2}y_n\left(-2\right)\right]-\left(y_{n+1}x+n_{c_1}y_n 1\right)+m^2y_n=0$$

$$\Rightarrow y_{n+2}\left(1-x^2\right)-y_{n+1}2nx-n\left(n-1\right)y_n-\left(y_{n+1}x+ny_n\right)+m^2y_n=0$$

$$\Rightarrow\left(1-x^2\right)y_{n+2}=\left(2n+1\right)xy_{n+1}+\left(n^2-m^2\right)y_n$$

Putting x = 0 in equations $y=\sin\left(m\sin^{-1}x\right)$, $\Rightarrow y_1=\frac{m}{\sqrt{1-x^2}}\cos\left(m\sin^{-1}x\right)$ and $\left(1-x^2\right)y_1^2=m^2\left(1-y^2\right)$

$$y\left(0\right)=0, y_1\left(0\right)=m \text{ and } y_2\left(0\right)=0$$

Putting x = 0 in above equation,

$$\left[y_{n+2}\left(1-x^2\right)+n_{c_1}y_{n+1}\left(-2x\right)+n_{c_2}y_n\left(-2\right)\right]-\left(y_{n+1}x+n_{c_1}y_n 1\right)+m^2y_n=0$$

$$\Rightarrow y_{n+2}\left(1-x^2\right)-y_{n+1}2nx-n\left(n-1\right)y_n-\left(y_{n+1}x+ny_n\right)+m^2y_n=0$$

$$\Rightarrow\left(1-x^2\right)y_{n+2}=\left(2n+1\right)xy_{n+1}+\left(n^2-m^2\right)y_n$$

$$y_{n+2}\left(0\right)=\left(n^2-m^2\right)y_n\left(0\right)$$

Putting n = 1,2,3 .. ... in the above equation, we get,

$$y_3\left(0\right)=\left(1^2-m^2\right)y_1\left(0\right)$$

$$=\left(1^2-m^2\right)m \quad \because y_1\left(0\right)=m$$

$$y_4\left(0\right)=\left(2^2-m^2\right)y_2\left(0\right)$$

$$=0 \qquad \because y_2\left(0\right)=0$$

$$y_5\left(0\right)=\left(3^2-m^2\right)y_3\left(0\right)$$

$$=m\left(1^2-m^2\right)\left(3^2-m^2\right)$$

$$\vdots$$

$$\Rightarrow y_n\left(0\right)=\begin{cases}0, & \text{if } n \text{ is even}\\ m\left(1^2-m^2\right)\left(3^2-m^2\right)\ldots\left[\left(n-2\right)^2-m^2\right], & \text{if } n \text{ is odd}\end{cases}$$

Example: If $y=e^{m\sin^{-1}x}$, show that, $\left(1-x^2\right)y_{n+2}-\left(2n+1\right)xy_{n+1}-\left(n^2+m^2\right)y_n=0$. Also find $y\ (\ )$.

Solution: Here $y=e^{m\sin^{-1}x}$,

$$\Rightarrow y_1=\frac{m}{\sqrt{1-x^2}}e^{m\sin^{-1}x}$$

$$=\frac{my}{\sqrt{1-x^2}}$$

$$\Rightarrow\left(1-x^2\right)y_1^2=m^2y^2$$

Differentiating above equation w.r.t. x, we get,

$$\left(1-x^2\right)2y_1y_2+y_1^2\left(-2x\right)=m^2 2yy_1$$

$$\Rightarrow\left(1-x^2\right)y_2-xy_1-m^2y=0$$

Differentiating above equation n times w.r.t. x using Leibnitz's theorem, we get,

$$\left[y_{n+2}\left(1-x^2\right)+n_{c_1}y_{n+1}\left(-2x\right)+n_{c_2}y_n\left(-2\right)\right]-\left(y_{n+1}x+n_{c_1}y_n1\right)-m^2y_n=0$$

$$\Rightarrow y_{n+2}\left(1-x^2\right)-y_{n+1}2nx-n\left(n-1\right)y_n-\left(y_{n+1}x+ny_n\right)-m^2y_n=0$$

$$\Rightarrow\left(1-x^2\right)y_{n+2}-\left(2n+1\right)xy_{n+1}-\left(n^2+m^2\right)y_n=0$$

To find $y_n(0)$: Putting x = 0 in equations $y=e^{m\sin^{-1}x}$, $\Rightarrow y_1=\frac{m}{\sqrt{1-x^2}}e^{m\sin^{-1}x}=\frac{my}{\sqrt{1-x^2}}$, and

$$\left(1-x^2\right)2y_1y_2+y_1^2\left(-2x\right)=m^2 2yy_1\Rightarrow\left(1-x^2\right)y_2-xy_1-m^2y=0$$

$$y\left(0\right)=1, y_1\left(0\right)=m \text{ and } y_2\left(0\right)=m^2$$

Also putting x = 0 we get,

$$y_{n+2}\left(0\right)=\left(n^2+m^2\right)y_n\left(0\right)$$

Putting n = 1,2,3 ... in the above equation, we get,

$$y_3\left(0\right)=\left(1^2+m^2\right)y_1\left(0\right)$$

$$=\left(1^2+m^2\right)m \qquad \because y_1\left(0\right)=m$$

$$y_4\left(0\right)=\left(2^2+m^2\right)y_2\left(0\right)$$

$$=m^2\left(2^2+m^2\right) \qquad \because y_2\left(0\right)=m^2$$

$$y_5\left(0\right)=\left(3^2+m^2\right)y_3\left(0\right)$$

$$=m\left(1^2+m^2\right)\left(3^2+m^2\right)$$

$$\vdots$$

$$\Rightarrow y_n\left(0\right)=\begin{cases} m^2\left(2^2+m^2\right)\ldots\left[\left(n-2\right)^2+m^2\right], & \text{if } n \text{ is even} \\ m\left(1^2+m^2\right)\left(3^2+m^2\right)\ldots\left[\left(n-2\right)^2+m^2\right], & \text{if } n \text{ is odd} \end{cases}$$

References

- Second-order-derivative, maths: byjus.com, Retrieved 01, January 2020
- Successive-Differentiation: magadhmahilacollege.org, Retrieved 16, June 2020
- Higher-order-derivatives, calculus: cliffsnotes.com, Retrieved 29, April 2020

- What-are-higher-order-derivatives, calculus: mathwarehouse.com, Retrieved 03, July 2020
- Leibniz-formula: math24.net, Retrieved 26, March 2020
- Chapter-1-Successive-Differentiation: theengineeringmaths.com, Retrieved 13, June 2020

4

Integral Calculus

The study of definitions, properties and applications of definite integral and indefinite integral are studied within the field of integral calculus. The process through which the value of an integral is determined is called integration. All the diverse principles of integral calculus such as line integral and Reimann integral have been carefully analyzed in this chapter.

The integration denotes the summation of discrete data. The integral is calculated to find the functions which will describe the area, displacement, volume, that occurs due to a collection of small data, which cannot be measured singularly. In a broad sense, in calculus, the idea of limit is used where algebra and geometry are implemented. Limits help us in the study of the result of points on a graph such as how they get closer to each other until their distance is almost zero. The concept of integration has developed to solve the following types of problems:

- To find the problem function, when its derivatives are given.
- To find the area bounded by the graph of a function under certain constraints.

These two problems lead to the development of the concept called the "Integral Calculus", which consist of definite and indefinite integral. In calculus, the concept of differentiating a function and integrating a function is linked using the theorem called the Fundamental Theorem of Calculus.

In Math, integration is a method of adding or summing up the parts to find the whole. It is a reverse process of differentiation, where we reduce the functions into parts. This method is used to find the summation under a vast scale. Calculation of small addition problems is an easy task which we can do manually or by using calculators as well. But for big addition problems, where the limits could reach to even infinity, integration methods are used. Integration and differentiation both are important parts of calculus. The concept level of these topics is very high. Hence, it is introduced to us at higher secondary classes and then in engineering or higher education.

Integral Calculus

Integral is based on a limiting procedure which approximates the area of a curvilinear region by breaking the region into thin vertical slabs. Let us now try to understand what does that mean: Take an example of a slope of a line in a graph to see what differential calculus is: In general, we can find the slope by using the slope formula. But what if we are given to find an area of a curve? For a curve, the slope of the points varies, and it is then we need differential calculus to find the slope of a curve.

Integration – Inverse Process of Differentiation

Differentiation is the process of finding the derivative of the functions and integration is the process of finding the antiderivative of a function. So, these processes are inverse of each other. Integration is the inverse process of differentiation or vice versa. The integration is also called the anti-differentiation. In this process, we are provided with the derivative of a function and asked to find out the function (i.e., primitive).

The differentiation of sin x is cos x. It is mathematically written as:

$$(d/dx)\ \sin x = \cos x$$

Here, cos x is the derivative of sin x. So, sin x is the antiderivative of the function cos x. Also, any real number "C" is considered as a constant function and the derivative of the constant function is zero. So, above equation can be written as,

$$(d/dx)\ (\sin x + C) = \cos x + 0$$

$$(d/dx)(\sin x + C) = \cos x$$

Where "C" is the arbitrary constant or constant of integration.

Generally, we can write the function as,

$(d/dx)[F(x)+C] = f(x)$, where x belongs to the interval I.

To represent the antiderivative of "f", the integral symbol "∫" symbol is introduced. The antiderivative of the function is represented as $\int f(x)\,dx$. This can also be read as the indefinite integral of the function "f" with respect to x. Therefore, the symbolic representation of the antiderivative of a function (Integration) is:

$$y = \int f(x)dx$$

$$\int f(x)dx = F(x) + C.$$

Integrals in Maths

You will come across, two types of integrals in maths:

- Definite Integral: An integral that contains the upper and lower limits then it is a definite integral. On a real line, x is restricted to lie. Riemann Integral is the other name of the Definite Integral. A definite Integral is represented as: $\int_a^b f(x)dx$.
- Indefinite Integral: Indefinite integrals are defined without upper and lower limits. It is represented as: $\int f(x)dx = F(x) + C$, Where C is any constant and the function f(x) is called the integrand.

Formulas for Integration

1. $\int 1dx = x + c$
2. $\int adx = ax + c$ where a is any constant.

3. $\int x^n dx = \frac{x^{n+1}}{n+1} + c$

4. $\int [f(x)]^n f'(x) dx = \frac{[f(x)]^{n+1}}{n+1} + c$

5. $\int \frac{1}{x} dx = \ln x + c$

6. $\int \frac{f'(x)}{f(x)} dx = \ln|f(x)| + c$

7. $\int a^x dx = \frac{a^x}{\ln a} + c$

8. $\int a^{f(x)} dx = \frac{a^{f(x)}}{\ln a} + c$

9. $\int e^x dx = e^x + c$

10. $\int e^{f(x)} dx = e^{f(x)} + c$

11. $\int af(x) dx = a \int f(x)$

12. $\int [f(x) \pm g(x)] dx = \int f(x) dx \pm \int g(x) dx$

13. $\int f(x) \cdot g(x) dx = f(x)\left(\int g(x) dx\right) - \int \left[f'(x)\left(\int g(x) dx\right)\right] dx$

14. $\int \ln x dx = x(\ln x - 1) + c$

15. $\int \sin x dx = \cos x + c$

16. $\int \cos x dx = \sin x + c$

17. $\int \tan x dx = \ln \sec x + c$

or

$-\ln \cos x + c$

18. $\int \cot x dx = \ln \sin x + c$

19. $\int \sec x dx = \ln(\sec x + \tan x) + c$

or

$\ln \tan\left(\frac{x}{2} + \frac{\pi}{4}\right) + c$

20. $\int \csc x dx = \ln(\csc x - \cot x) + c$

or

$\ln \tan \frac{x}{2} + c$

21. $\int \sec^2 x dx = \tan x + c$

22. $\int \csc^2 x dx = -\cot x + c$

23. $\int \sec x \tan x dx = \sec x + c$

24. $\int \csc\ x \cot x dx = -\csc x + c$

25. $\int \sinh x dx = \cosh x + c$

26. $\int \cosh x dx = \sinh x + c$

27. $\int \tanh x dx = \ln \cosh x + c$

28. $\int \coth x dx = \ln \sinh x + c$

29. $\int \sec hx dx = \tan^{-1}(\sinh x) + c$

30. $\int \csc hx dx = -\coth^{-1}(\cosh x)$

31. $\int \sec h^2 x dx = \tanh x + c$

32. $\int \operatorname{csh} h^2 x dx = -\coth x + c$

33. $\int \sec hx \tanh x dx = -\sec hx + c$

34. $\int \csc hx \coth x dx = -\csc hx + c$

35. $\int \frac{1}{\sqrt{a^2 - x^2}} dx = \sin^{-1} \frac{x}{a} + c$

or

$\cos^{-1} \frac{x}{a} + c$

36. $\int \frac{1}{\sqrt{x^2 - a^2}} dx = \cosh^{-1} \frac{x}{a} + c$

or

$\ln\left(x + \sqrt{x^2 - a^2}\right) + c$

37. $\int \frac{1}{\sqrt{x^2+a^2}}dx = \sinh^{-1}\frac{x}{a}+c$

or

$\ln\left(x+\sqrt{x^2+a^2}\right)+c$

38. $\int \frac{1}{a^2-x^2}dx = \frac{1}{a}\tanh^{-1}\frac{x}{a}+c$

or

$\frac{1}{2a}\ln\left(\frac{a+x}{a-x}\right)+c$

39. $\int \frac{1}{x^2-a^2}dx = -\frac{1}{a}\coth^{-1}\frac{x}{a}+c$

or

$\frac{1}{2a}\ln\left(\frac{x-a}{x+a}\right)+c$

40. $\int \frac{1}{x^2+a^2}dx = \frac{1}{a}\tan^{-1}\frac{x}{a}+c$

41. $\int \frac{1}{x\sqrt{a^2-x^2}}dx = -\frac{1}{a}\text{sech}^{-1}\frac{x}{a}+c$

or

$-\frac{1}{a}\ln\left(\frac{a+\sqrt{a^2-x^2}}{x}\right)+c$

42. $\int \frac{1}{x\sqrt{x^2-a^2}}dx = \frac{1}{a}\sec^{-1}\frac{x}{a}+c$

43. $\int \frac{1}{x\sqrt{x^2+a^2}}dx = -\frac{1}{a}\text{csch}^{-1}\frac{x}{a}+c$

or

$-\frac{1}{a}\ln\left(\frac{a+\sqrt{x^2+a^2}}{x}\right)+c$

44. $\int \sqrt{a^2-x^2}\,dx = \frac{1}{2}x\sqrt{a^2-x^2}+\frac{a^2}{2}\sin^{-1}\frac{x}{a}+c$

45. $\int \sqrt{x^2-a^2}\,dx = \frac{1}{2}x\sqrt{x^2-a^2}-\frac{a^2}{2}\cosh^{-1}\frac{x}{a}+c$

or

$\frac{1}{2}x\sqrt{x^2-a^2}-\frac{a^2}{2}\ln\left(x+\sqrt{x^2-a^2}\right)+c$

46. $\int \sqrt{x^2+a^2}\,dx = \frac{1}{2}x\sqrt{x^2+a^2} + \frac{a^2}{2}\sinh^{-1}\frac{x}{a} + c$

or

$\frac{1}{2}x\sqrt{x^2+a^2} + \frac{a^2}{2}\ln\left(x+\sqrt{x^2+a^2}\right) + c$

47. $\int e^{ax}\sin(bx+c)\,dx = \frac{e^{ax}}{a^2+b^2}\left[a\sin(bx+c) - b\cos(bx+c)\right]$

48. $\int e^{ax}\cos(bx+c)\,dx = \frac{e^{ax}}{a^2+b^2}\left[a\cos(bx+c) + b\sin(bx+c)\right]$

49. $\int \sin mx\cos nx\,dx = \frac{\cos(m+n)x}{2(m+n)} - \frac{\cos(m-n)x}{2(m-n)} + c$

50. $\int \sin mx\sin nx\,dx = \frac{\sin(m+n)x}{2(m+n)} + \frac{\sin(m-n)x}{2(m-n)} + c$

51. $\int \cos mx\cos nx\,dx = \frac{\sin(m+n)x}{2(m+n)} + \frac{\sin(m-n)x}{2(m-n)} + c$

52. $\int \sin^{-1}x\,dx = x\sin^{-1}x + \sqrt{1-x^2} + c$

53. $\int \cos^{-1}x\,dx = x\cos^{-1}x - \sqrt{1-x^2} + c$

54. $\int \tan^{-1}x\,dx = x\tan^{-1}x - \frac{1}{2}\ln\left(1+x^2\right) + c$

55. $\int \cot^{-1}x\,dx = x\cot^{-1}x + \frac{1}{2}\ln\left(1+x^2\right) + c$

56. $\int \sec^{-1}x\,dx = x\sec^{-1}x - \ln\left(x+\sqrt{x^2-1}\right) + c$

57. $\int \csc^{-1}x\,dx = x\csc^{-1}x + \ln\left(x+\sqrt{x^2-1}\right) + c$

58. $\int \frac{1}{a+b\sin x}\,dx = \frac{2}{\sqrt{a^2-b^2}}\tan^{-1}\left(\frac{a\tan^{-1}\frac{x}{2}+b}{\sqrt{a^2-b^2}}\right) + c$

if $a^2 > b^2$

59. $\int \frac{1}{a+b\sin x}\,dx = \frac{1}{\sqrt{a^2-b^2}}\ln\left(\frac{a\tan\frac{x}{a}+b-\sqrt{b^2-a^2}}{a\tan\frac{x}{a}+b+\sqrt{b^2-a^2}}\right) + c$

if $a^2 < b^2$

60. $\int \frac{1}{a + b\cos x} dx = \frac{2}{\sqrt{a^2 - b^2}} \tan^{-1}\left(\sqrt{\frac{a-b}{a+b}} \tan\frac{x}{2}\right) + c$

if $a^2 > b^2$

61. $\int \frac{1}{a + b\cos x} dx = \frac{1}{\sqrt{a^2 - b^2}} \ln\left(\frac{\sqrt{b+a} + \tan\frac{x}{2}\sqrt{b-a}}{\sqrt{b+a} - \tan\frac{x}{2}\sqrt{b-a}}\right) + c$

if $a^2 < b^2$

62. $\int \frac{1}{a + b\sinh x} dx = \frac{1}{\sqrt{a^2 + b^2}} \ln\left(\frac{\sqrt{a^2+b^2} + a\tanh\frac{x}{2} - b}{\sqrt{a^2+b^2} - a\tanh\frac{x}{2} + b}\right) + c$

63. $\int \frac{}{a \quad b\cosh x} dx = \frac{\sqrt{a+b} + \sqrt{a-b}\tanh -}{\sqrt{\quad +b} - \sqrt{a-b}\tanh -} +$

if $a \quad b$

64. $\int \frac{1}{a + b\cosh x} dx = \frac{2}{\sqrt{b^2 - a^2}} \tan^{-1}\sqrt{\frac{b-a}{b+a}} \tanh^{-1}\frac{x}{2} + c$

if $a > b$

Integration as Summation

The figure below on the left shows an area bounded by the x axis, the lines x = a and x = b, and the curve y = f(x). Note that the area lies entirely above the x axis.

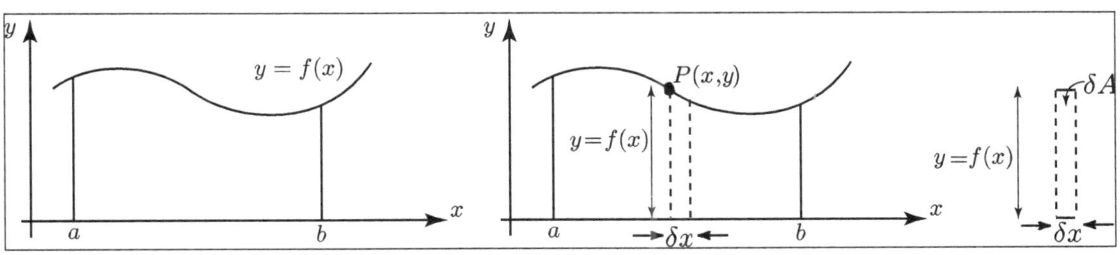

There are several ways in which this area can be estimated. Suppose we split the area into thin vertical strips, like the one shown, and treat each strip as being approximately rectangular. The sum of the areas of the rectangular strips then gives an approximate value for the area under the curve. The thinner the strips, the better will be the approximation. A typical strip is shown drawn from the point P(x, y). The width of the strip is labelled δx. We label it like this because the symbol δ is used to indicate a small increase in the variable being considered, in this case x. The height of the strip is equal to the y value on the curve at point P that is f(x). So the area of the strip shown is approximately f(x) δx. Suppose we let the area of this small strip be δA. We use the delta notation again, because this strip makes a small contribution, δA, to the total area, A, under the curve. Then,

$$\delta A \approx f(x)\delta x$$

Now if we add up the areas of all such thin strips from a to b, which we denote by $\Sigma_{x=a}^{b}\delta A$, we obtain the total area under the curve.

$$\text{total area} = \sum_{x=a}^{b}\delta A \approx \sum_{x=a}^{b} f(x)\delta x$$

To make this approximation more accurate we must let the thickness of each strip become very small indeed, that is, we let $\delta x \to 0$, giving,

$$\text{total area} = \lim_{\delta x \to 0}\sum_{x=a}^{b} f(x)\delta x$$

The notation $\lim_{\delta x \to 0}$ means that we consider what happens to the expression following it as δx gets smaller and smaller. This is known as the limit of a sum. If this limit exists we write it formally as,

$$\int_a^b f(x)dx$$

thus defining a definite integral as the limit of a sum. Thus we have the important result that,

$$\boxed{\int_a^b f(x)dx = \lim_{\delta x \to 0}\sum_{x=a}^{b} f(x)\delta x}$$

Integration can therefore be regarded as a process of adding up, that is as a summation. Whenever we wish to find areas under curves, volumes etc., we can do this by finding the area or volume of a small portion, and then summing over the whole region of interest. The calculation can then be performed using the technique of definite integration.

Example: Suppose a unit charge moves along a curve C in an electric field E. At any point on the curve the electric field vector can be resolved into two perpendicular components, E_t say, along the curve, and E_n perpendicular, or normal to the curve.

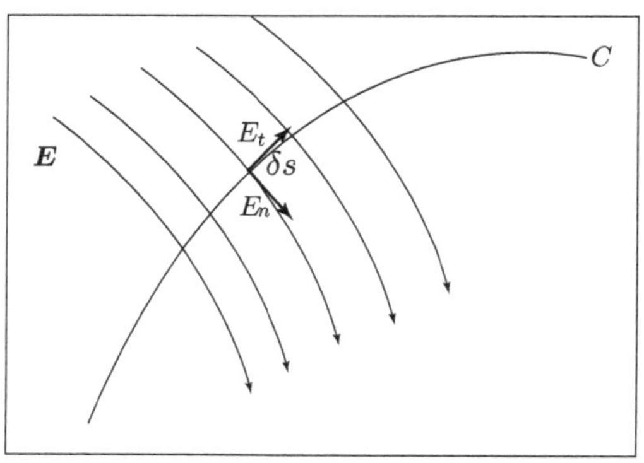

In moving the charge a small distance δs along the curve the electric field does work equal to $E_t\delta s$, because only the tangential component does work. To find the total work done as the charge moves along the length of the curve we must sum all such small contributions, i.e

$$\text{total work done} = \sum E_t \delta s, \quad \text{in the limit as } \delta s \to 0$$

that is,

$$\text{total work done} = \lim_{\delta s \to 0} \sum E_t \delta s$$

which defines the integral $\int_C E_t \delta s$. The symbol $\int_C$ tells us to sum the contributions along the curve C. This is an example of a line integral because we integrate along the line (curve) C.

Line Integral

Let $\gamma : [a, b] \to \mathbb{R}^n$ be a piecewise smooth curve, A is an open set in R^n containing the trace of the curve γ and $f : A \to \mathbb{R}^n$ be a bounded function. The line integral of f along γ is denoted by $\int \langle f, d\gamma \rangle$ and is defined by,

$$\int \langle f, d\gamma \rangle = \int_a^b \langle f(\gamma(t)), \gamma'(t) \rangle dt,$$

provided the integral exists (even as an improper integral).

In the above definition, $\langle \cdot, \cdot \rangle$ denotes the standard inner product on R^n.

Example: Let $a > 0$. Evaluate the integral $\int \langle f, d\gamma \rangle$ where $f : R^2 \to \cdot^2$ is defined by,

$$f(x, y) = (y, 2a - x)$$

and the path $\gamma : [0, 2\pi] \to \cdot^2$ is defined by,

$$\gamma(t) = (a(1 - \cos t), a(t - \sin t)), \text{ for all } t \in [0, 2\pi].$$

Solution: Given for all $t \in [0, 2\pi]$,

$$\gamma(t) = (a(1 - \cos t), a(t - \sin t)).$$

Differentiating with respect to t we find,

$$\gamma'(t) = (a \sin t, a - a \cos t).$$

So,

$$f(\gamma(t)) = f(a(1 - \cos t), a(t - \sin t)) = (at - a \sin t, a + a \cos t).$$

Consequently,

$$\begin{aligned}\langle f(\gamma(t)),\ \gamma'(t)\rangle &= \langle (at-a\sin t,\ a+a\cos t),(a\sin t,\ a-a\cos t)\rangle\\ &= a^2\langle (t-\sin t,\ 1+\cos t),(\sin t,\ 1-\cos t)\rangle\\ &= a^2\{t\sin t-\sin^2 t+1-\cos^2 t\}\\ &= a^2\ t\sin t.\end{aligned}$$

Therefore,

$$\int\langle f,d\gamma\rangle=\int_0^{2\pi}a^2 t\sin t\,dt=a^2\int_0^{2\pi}t\sin t\,dt.$$

Using integration by parts we find,

$$\int t\sin t\,dt=-t\cos t+\sin t.$$

So, $\int\langle f,d\gamma\rangle=a^2\int_0^{2\pi}t\sin t\,dt=-2\pi a^2.$

In some cases we have to parameterise the given curve.

Example: Let a, b > 0. Evaluate the line integral of the vector field $f: R^2 \to \sim^2$ defined by,

$$f(x,y)=(px+qy,\ px-qy)\ \text{for all}\ (x,\ y)\in\mathbb{R}^2$$

once around the ellipse,

$$\frac{x^2}{a}+\frac{y^2}{b^2}=1$$

in the anticlockwise direction. Here p and q are two arbitrary constants.

Solution: We can parameterise the given curve as,

$$\gamma(t)=(x(t),\ y(t))=(a\cos t, b\sin t),\ \text{for all}\ t\in[0,\ 2\pi].$$

Then for all $t\in[0,\ 2\pi]$,

$$\gamma'(t)=(x'(t),\ y'(t))=(-a\sin t,\ b\cos t),$$

accordingly the curve under consideration is smooth.

Now, for all $t\in[0,\ 2\pi]$,

$$\begin{aligned}f(\gamma(t))&=(px(t)+qy(t),\ px(t)-qy(t))\\ &=(pa\cos t+qb\sin t,\ pa\cos t-qb\sin t).\end{aligned}$$

So,

$$\langle f(\gamma(t)),\gamma'(t)\rangle = (pa\cos t + qb\sin t)(-a\sin t) + (pa\cos t - qb\sin t)(b\cos t)$$
$$= pab\cos^2 t - qab\sin^2 t - (pa^2 + qb^2)\sin t\cos t.$$

A little calculation yields,

$$\int \langle f(\gamma(t)),\ \gamma'(t)\rangle dt$$
$$= \frac{1}{2}pab\left(t + \frac{\sin 2t}{2}\right) - \frac{1}{2}qab\left(t - \frac{\sin 2t}{2}\right) - \left(\frac{pa^2 + qb^2}{2}\right)\left(-\frac{\cos 2t}{2}\right),$$

ignoring the arbitrary constant for definite integration.

Therefore,

$$\int \langle f, d\gamma\rangle = \int_0^{2\pi} \langle f(\gamma(t)),\gamma'(t)\rangle dt = (p-q)ab\pi.$$

An inquisitive reader may doubt whether all the parameterisations would yield the same result. A line integral remains unaltered under change of parameter provided the orientation remains the same.

Let $\gamma:[a,b]\to\mathbb{R}^m$ be a smooth curve, $A\subset R^n$ be an open set containing the trace of γ and $f: A\to\mathbb{R}^m$ be a vector field. If we express both f and γ in terms of their component functions, i.e.

$$f = (f_1, f_2, \ldots, f_m),\ \gamma = (\gamma_1, \gamma_2, \ldots, \gamma_m);$$

then it follows that,

$$\langle f(\gamma(t)),\gamma'(t)\rangle$$
$$= \langle (f_1(\gamma(t)), f_2(\gamma(t)), \ldots, f_m(\gamma(t))), (\gamma_1'(t), \gamma_2'(t), \ldots, \gamma_m'(t))\rangle$$
$$= \sum_{k=1}^{m} f_k(\gamma(t))\gamma_k'(t).$$

So,

$$\int_a^b \langle f, d\gamma\rangle = \int_a^b \sum_{k=1}^{m} f_k(\gamma(t))\gamma_k'(t)dt$$
$$= \sum_{k=1}^{m}\int_a^b f_k(\gamma(t))\gamma_k'(t)dt$$
$$= \sum_{k=1}^{m}\int_a^b \langle f_k, d\gamma_k\rangle.$$

This enables us to write,

$$\int_a^b \langle f, d\gamma\rangle = \int_a^b f_1 d\gamma_1 + \ldots + \int_a^b f_m d\gamma_m.$$

It is to be noticed that all $f_1, \ldots, f_m, \gamma_1, \ldots, \gamma_m$ are real valued functions. So their usual product coincides with their inner product.

Example: Let us consider the line integral,

$$\int_C \frac{(x-y)dx+(x+y)dy}{\left(x^2+y^2\right)^2},$$

where C is the circle given by $x^2+y^2=a^2$, $a>0$.

Here the integrand is given by,

$$f(x, y)=\left(\frac{x-y}{\left(x^2+y^2\right)^2}, \frac{x+y}{\left(x^2+y^2\right)^2}\right)=\left(f_1(x,y), f_2(x,y)\right), \text{ say.}$$

Clearly, it is defined anywhere in R^2 except at the origin. We are to find the line integral of this vector field with respect to the circle $x^2+y^2=a^2$, $a>0$. A parametric form for this circle is given by, $x=a\cos t, y=a\sin t$.

We find,

$$\int_C\left(\frac{x-y}{\left(x^2+y^2\right)^2}, \frac{x+y}{\left(x^2+y^2\right)^2}\right)dC$$

$$=\int_C \frac{x-y}{\left(x^2+y^2\right)^2}dx+\int_C \frac{x+y}{\left(x^2+y^2\right)^2}dy.$$

Putting x = a cos t and y = a sin t the required integral becomes,

$$\int_0^{2\pi} \frac{a\cos t-a\sin t}{\left(a\cos^2 t+a\sin^2 t\right)^2}d(a\cos t)+\int_0^{2\pi} \frac{a\cos t+a\sin t}{\left(a\cos^2 t+a\sin^2 t\right)^2}d(a\sin t)$$

$$=\int_0^{2\pi}(\cos t-\sin t)(-\sin t dt)+\int_0^{2\pi}(\cos t+\sin t)(\cos t dt)$$

$$=\int_0^{2\pi}dt$$

$$=2\pi.$$

Example: Let us consider the vector field f defined as follows:

$$f(x,y):=\left(x+\sqrt{y}, x-\sqrt{y}\right), \text{ for all } x,y\in\mathbb{R},\ y\geq 0.$$

Here we are considering the non-negative square root only. Let us find the line integral of f from the origin to the point (1, 1):

- First along the line segment joining these two points.

- Then along the curve α defined by, $\alpha(t):=(t^3,t^2)$.

Solution: Let γ be the straight line segment in R^2 joining the points (0, 0) and (1, 1). Then we can express,

$\gamma(t):=(t,t)$ for all $t\in[0,1]$.

Clearly both the curves γ and α are smooth. Now,

$$f(\gamma(t))=f(t,t)=(t+\sqrt{t},t-\sqrt{t})$$
$$\Rightarrow\langle f(\gamma(t)),\gamma'(t)\rangle=\langle(t+\sqrt{t},\ t-\sqrt{t}),(1,\ 1)\rangle$$
$$2t.$$

Hence, $\int_0^1\langle f,d\gamma\rangle=\int_0^1\langle f(\gamma(t)),\gamma'(t)\rangle dt=\int_0^1 2tdt=1.$

For the curve α we find,

$$f(\alpha(t))=f(t^3,\ t^2)=(t^3+t,\ t^3-t).$$

So, $\langle f(\alpha(t)),\ \alpha'(t)\rangle=\langle(t^3+t,\ t^3-t),(3t^2,\ 2t)\rangle=3t^5+2t^4+3t^3-2t^2.$

Hence,

$$\begin{aligned}\int_0^1\langle f,d\gamma\rangle&=\int_0^1\langle f(\gamma(t)),\ \gamma'(t)\rangle dt\\&=\int_0^1(3t^5+2t^4+3t^3-2t^2)dt\\&=\frac{59}{60}.\end{aligned}$$

This example shows that the value of a line integral (if exists) depends not on the initial and terminal points of the curve only, but on the path joining them also. Now we are going to discuss some fundamental properties of line integral.

Result: Line integral is linear with respect to the integrand. That is, let $\gamma:[a,\ b]\to\mathbb{R}^n$ be a piecewise smooth curve and f, $g:A\to\mathbb{R}^n$ are two functions such that both $\int_a^b\langle f,d\gamma\rangle$ and $\int_a^b\langle g,d\gamma\rangle$ exists, here trace $\gamma\subset A\subset\mathbb{R}^n$. Then for every $\lambda,\mu\in\mathbb{R},\int_a^b\langle\lambda f+\mu g,d\gamma\rangle$ exists and

$$\int_a^b\langle\lambda f+\mu g,d\gamma\rangle=\lambda\int_a^b\langle f,d\gamma\rangle+\mu\int_a^b\langle g,d\gamma\rangle.$$

Proof: We have, for all t ∈ [a, b]

$$\begin{aligned}(\lambda f+\mu g)(\gamma(t)) &= (\lambda f)(\gamma(t))+(\mu g)(\gamma(t))\\ &= \lambda f(\gamma(t))+\mu g(\gamma(t))\end{aligned}$$

$$\begin{aligned}\Rightarrow \langle(\lambda f+\mu g)(\gamma(t)),\ \gamma'(t)\rangle &= \langle\lambda f(\gamma(t))+\mu g(\gamma(t)),\ \gamma'(t)\rangle\\ &= \lambda\langle f(\gamma(t)),\ \gamma'(t)\rangle+\mu\langle g(\gamma(t)),\ \gamma'(t)\rangle.\end{aligned}$$

Since both $\int_a^b \langle f,d\gamma\rangle$ and $\int_a^b \langle g,d\gamma\rangle$ exists, from the above relation it follows that,

$$\begin{aligned}&\int_a^b \langle(\lambda f+\mu g)(\gamma(t)),\ \gamma'(t)\rangle dt\\ &= \int_a^b \{\lambda\langle f(\gamma(t)),\gamma'(t)\rangle+\mu\langle g(\gamma(t)),\ \gamma'(t)\rangle\} dt\\ &= \lambda\int_a^b \langle f(\gamma(t)),\gamma'(t)\rangle dt+\mu\int_a^b \langle g(\gamma(t)),\ \gamma'(t)\rangle dt.\end{aligned}$$

This completes the proof.

Join of Two Curves: Let $\gamma_1:[a,b]\to\mathbb{R}^n$ and $\gamma_2:[b,c]\to\mathbb{R}^n$ be two curves satisfying $\gamma_1(b) = \gamma_2(b)$. We define a new curve $\gamma:[a,c]\to\mathbb{R}^n$ as follows:

$$\begin{aligned}\gamma(t) &= \gamma_1(t),\ \text{if } t\in[a,b],\\ &= \gamma_2(t),\ \text{if } t\in[b,c].\end{aligned}$$

(Clearly, the continuity of γ at any point in $[a, c]\backslash\{b\}$ follows from the continuity of γ_1 or γ_2. Its continuity at the point c follows from the pasting lemma.) Thus γ is a curve in R^n called 'the join of γ_1 and γ_2'. This new curve γ is usually denoted by $\gamma_1 + \gamma_2$. Two points are to be noted here: First we can join finitely many curves in this manner. Second, this join is associative.

Result: Let $\gamma_1:[a,b]\to\mathbb{R}^n$ and $\gamma_2:[b,c]\to\mathbb{R}^n$ be two piecewise smooth curves and $\gamma = \gamma_1 + \gamma_2$. Also let A be an open set in R^n containing the trace of γ and $f: A\to\mathbb{R}^n$ be a bounded function such that both the line integrals $\int_a^b \langle f,d\gamma_1\rangle$ and $\int_b^c \langle f,d\gamma_2\rangle$ exist. Then $\int_a^c \langle f,d\gamma\rangle$ exists and

$$\int_a^c \langle f,d\gamma\rangle = \int_a^b \langle f,d\gamma_1\rangle+\int_b^c \langle f,d\gamma_2\rangle$$

In other words,

$$\int_\gamma \langle f,d\gamma\rangle = \int_{\gamma_1+\gamma_2} \langle f,d\gamma\rangle = \int_{\gamma_1} \langle f,d\gamma_1\rangle+\int_{\gamma_2} \langle f,d\gamma_2\rangle.$$

Proof: From the given condition we find that,

$$\begin{aligned}\int_a^c \langle f,d\gamma\rangle &= \int_a^c \langle f(\gamma(t)),\gamma'(t)\rangle dt\\ &= \int_a^b \langle f(\gamma(t)),\gamma'(t)\rangle dt+\int_a^c \langle f(\gamma(t)),\gamma'(t)\rangle dt\\ &= \int_a^b \langle f(\gamma_1(t)),\gamma_1'(t)\rangle dt+\int_a^c \langle f(\gamma_2(t)),\gamma_2'(t)\rangle dt\\ &= \int_a^b \langle f,d\gamma_1\rangle+\int_b^c \langle f,d\gamma_2\rangle.\end{aligned}$$

Hence the result.

Example: Let a > 0. Calculate the line integral,

$$\int_\gamma \frac{dx+dy}{|x|+|y|},$$

where γ is the square with vertices A(0, −a), B(a, 0), C(0, a) and D(−a, 0), traversed once in the counterclockwise direction. We can write,

$$\gamma = \gamma_1 + \gamma_2 + \gamma_3 + \gamma_4,$$

where γ_1, γ_2, γ_3, γ_4 denote respectively the line segments AB, BC, CD and DA. We can parameterise γ_1, γ_2, γ_3, γ_4 one by one as follows:

$$\gamma_1(t) = (t, -a+t) \text{ for all } t \in [0,\ a],$$
$$\gamma_2(t) = (a-t, t) \text{ for all } t \in [0,\ a],$$
$$\gamma_3(t) = (-t,\ a-t) \text{ for all } t \in [0,\ a],$$
$$\gamma_4(t) = (-a+t, -t) \text{ for all } t \in [0,\ a].$$

Here the integrand is a vector field $f : \mathbb{R}^2 \setminus \{(0,0)\} \to \mathbb{R}^2$ defined by,

$$f(x,\ y) = \left(\frac{1}{|x|+|y|}, \frac{1}{|x|+|y|}\right) \text{for all} (x,\ y) \in \mathbb{R}^2 \setminus \{(0,0)\}.$$

The curve γ_1 lies in the fourth quadrant, in which the first coordinate is nonnegative and the second coordinate is non-positive. So, for points on γ_1,

$$f(x,\ y) = \left(\frac{1}{|x|+|y|}, \frac{1}{|x|+|y|}\right) = \left(\frac{1}{t+a-t}, \frac{1}{t+a-t}\right) = \left(\frac{1}{a}, \frac{1}{a}\right).$$

Consequently,

$$\int_{\gamma_1} \langle f, d\gamma_1 \rangle = \int_0^a \left\langle \left(\frac{1}{a}, \frac{1}{a}\right), (1,1) \right\rangle = \int_0^a \frac{2}{a} dt = 2.$$

In a similar process, one can check,

$$\int_{\gamma_2} \langle f, d\gamma_2 \rangle = \int_0^a \left\langle \left(\frac{1}{a}, \frac{1}{a}\right), (-1,1) \right\rangle = 0,$$

$$\int_{\gamma_3} \langle f, d\gamma_3 \rangle = \int_0^a \left\langle \left(\frac{1}{a}, \frac{1}{a}\right), (-1,-1) \right\rangle = -2,$$

$$\int_{\gamma_4} \langle f, d\gamma_4 \rangle = \int_0^a \left\langle \left(\frac{1}{a}, \frac{1}{a}\right), (1,-1) \right\rangle = 0.$$

Adding we find, $\int_\gamma \langle f, d\gamma \rangle = 0$.

Reparameterisation of a Curve and its Effect on Line Integral

Let $\gamma:[a,b]\to\mathbb{R}^n$ be a smooth curve and $h:[c,d]\to[a,b]$ be a continuous function. Then $\beta=\gamma\circ h:[c,d]\to\mathbb{R}^n$ is another curve in R^n, called a reparameterisation of γ by h. Assume that the change of parameter function h satisfies the following two conditions:

- $h:[c,d]\to[a,b]$, is bijective.
- h is continuously differentiable over [c, d] with $h'(s)\neq 0$ for all s ∈ [c, d].

This kind of change of parameter functions induce an equivalence relation '~' over all smooth curves in R^n sharing the same trace. We define $\gamma \sim \alpha$ if and only if α is such a reparameterisation of γ.

Let $\gamma:[a,b]\to\mathbb{R}^n$ be an arbitrary smooth curve. The identity map $i:[a,b]\to[a,b]$ satisfies both the above conditions mentioned above. Since $\gamma = \gamma$ o i, it follows that $\gamma \sim \gamma$. Since γ was an arbitrarily chosen smooth curve in R^n, it follows that the relation '~' is reflexive.

Next let α and γ be two curves in R^n satisfying $\gamma \sim \alpha$. Then there exists a map $h:[c,d]\to[a,b]$ satisfying both the above mentioned conditions, i.e. h is bijective and continuously differentiable with nowhere vanishing derivative, where $\alpha = \gamma \circ h$. By Inverse Function Theorem, h^{-1} exists, is continuously differentiable and for all s ∈ [c, d],

$$(h^{-1})'(h(s))=\frac{1}{h'(s)}.$$

Thus h^{-1} also satisfies the given two conditions and $\gamma = \alpha \circ h^{-1}$. Consequently, $\alpha \sim \gamma$. Therefore the relation '~' is symmetric.

Finally let $\gamma \sim \alpha$ and $\alpha \sim \beta$. Then there exists two continuously differentiable bijective functions h_1, h_2 with nowhere vanishing derivative and $\alpha = \gamma \circ h_1$, $\beta = \alpha \circ h_2$. Let $h = h_1 \circ h_2$. Then $\beta = \gamma \circ h$. Since the composition of two bijective functions is bijective, it follows that h is bijective. Chain rule assists us to convince that h is continuously differentiable and h o vanishes nowhere in our scope of discussion. $\gamma \sim \beta$. Thus the relation '~' is transitive and hence an equivalence relation.

Result: Let γ and α be two equivalent smooth curves in R^n and f is a bounded function defined over an open set containing the trace of these two curves (note that they have common trace). If the change of parameter function h is orientation preserving, then,

$$\int\langle f,d\gamma\rangle=\int\langle f,d\alpha\rangle.$$

Proof: Given $\gamma:[a,b]\to\mathbb{R}^n$ be a smooth curve, $h:[c,d]\to[a,b]$ is a bijective continuously differentiable function with nowhere vanishing derivative and $\alpha = \gamma \circ h$. Then by chain rule, for all s ∈ [c, d],

$$\alpha'(s)=(\gamma\circ h)'(s)=\gamma'(h(s))h'(s)=\gamma'(t)h'(s).$$

So, for all s ∈ [c, d],

$$\langle f(\alpha(s)),\ \alpha'(s)\rangle = \langle f(\gamma(t)),\ \gamma'(t)h'(s)\rangle = h'(s)\langle f(\gamma(t)),\ \gamma'(t)\rangle.$$

Therefore,

$$\begin{aligned}\int_c^d \langle f(\alpha(s)),\ \alpha'(s)\rangle ds &= \int_c^d h'(s)\langle f(\gamma(t)),\ \gamma'(t)\rangle ds \\ &= \int_{h(c)}^{h(d)} \langle f(\gamma(t)),\ \gamma'(t)\rangle dt \\ &\quad \text{substituting t for s,} \\ &= \int_a^b \langle f(\gamma(t)),\ \gamma'(t)\rangle dt,\end{aligned}$$

since h is orientation preserving, h(c) = a, h(d) = b. Hence the result follows.

If α would be an orientation reversing reparameterisation of γ, then the proof would involve similar steps. There would be no change in the integrand. Only at the last step, while substituting t for s, we find,

$$\begin{aligned}\int_c^d \langle f(\alpha(s)),\ \alpha'(s)\rangle ds &= \int_{h(c)}^{h(d)} \langle f(\gamma(t)),\ \gamma'(t)\rangle dt \\ &= \int_b^a \langle f(\gamma(t)),\ \gamma'(t)\rangle dt \\ &= -\int_a^b \langle f(\gamma(t)),\ \gamma'(t)\rangle dt.\end{aligned}$$

Thus the value of the line integral would be the same, with a change in sign. In case we consider γ a piecewise smooth curve, we can arrive at similar steps as shown above. The only difference is that we would perform the same steps for each 'piece' in which the curve is smooth.

Result: Let $\gamma:[a,b]\to\mathbb{R}^n$ be a piecewise smooth curve, A is an open set in R^n containing the trace of the curve γ and $f: A \to$ be a bounded function. Let the arc length of γ be L and M be an upper bound for $\|f\|$ in A. Then we observe that,

$$\begin{aligned}\int\langle f, d\gamma\rangle &= \int_a^b \langle f,(\gamma(t)),\gamma'(t)\rangle dt \\ &\le \int_a^b \left|\langle f,(\gamma(t)),\gamma'(t)\rangle\right| dt \\ &\le \int_a^b \|f(\gamma(t))\|\ \|\gamma'(t)\| dt \\ &\le M\int_a^b \|\gamma'(t)\| dt \\ &= ML.\end{aligned}$$

This result is known as ML-formula. This gives us an upper bound of a line integral.

Line Integral with Respect to Arc Length Parameter

Let $\gamma:[a,b]\to\mathbb{R}^n$ be a smooth curve. Then the arc length function of γ is given by,

$$s(t) = \int_a^t \|\gamma'(u)\| du.$$

Then s is differentiable and for all $t \in [a,b]$, $s'(t) = \|\gamma'(t)\|$.

Now let $f : A \to \mathbb{R}^n$ be a bounded function, where A is an open set in R^n containing trace of γ. Let us define a scalar field F on the trace of γ as follows:

$$F(\gamma(t)) := \left\langle f(\gamma(t)), \frac{\gamma'(t)}{\|\gamma'(t)\|} \right\rangle \text{ for all } t \in [a,b].$$

Then,

$$F(\gamma(t))s'(t) = \langle f(\gamma(t)), \gamma'(t) \rangle$$

$$\Rightarrow \int_a^b F(\gamma(t))s'(t)dt = \int_a^b \langle f(\gamma(t)), \gamma'(t) \rangle dt = \int \langle f, d\gamma \rangle.$$

We can write ds for s'(t)dt and change the limits of integration accordingly. This motivates us to define the line integral of a scalar field defined on the trace of a rectifiable curve with respect to the arc length parameter.

Let F be a scalar field defined on the trace of a rectifiable curve γ and s denote the arc length function of γ (starting from a suitable point). Then,

$$\int_\gamma F ds = \int_a^b F(\gamma(t))s'(t)dt.$$

Riemann Integral

The Riemann integral is the definite integral normally encountered in calculus texts and used by physicists and engineers. Other types of integrals exist (e.g., the Lebesgue integral), but are unlikely to be encountered outside the confines of advanced mathematics texts. Cases where these methods [i.e., generalizations of the Riemann integral] are applicable and Riemann's [definition of the integral] is not are too rare in physics to repay the extra difficulty. The Riemann integral is based on the Jordan measure, and defined by taking a limit of a Riemann sum,

$$\int_a^b f(x)dx \equiv \lim_{\max \Delta x_k \to 0} \sum_{k=1}^n f(x_k^*) \Delta x_k$$

$$\iint f(x,y)dA \equiv \lim_{\max \Delta A_k \to 0} \sum_{k=1}^n f(x_k^*, y_k^*) \Delta A_k$$

$$\iiint f(x,y,z)dV \equiv \lim_{\max \Delta V_k \to 0} \sum_{k=1}^n f(x_k^*, y_k^*, z_k^*) \Delta V_k$$

where $a \le x \le b$ and x_k^*, y_k^*, and z_k^* are arbitrary points in the intervals Δx_k, Δy_k, and Δz_k, respectively. The value max Δx_k is called the mesh size of a partition of the interval [a, b] into subintervals Δx_k.

As an example of the application of the Riemann integral definition, find the area under the curve $y = x^r$ from 0 to a. Divide (0, a) into n segments, so $\Delta x_k = a/n \equiv h,$ then,

$$f(x_1) = f(0) = 0$$
$$f(x_2) = f(\Delta x_k) = h^r$$
$$f(x_3) = f(2\Delta x_k) = (2h)^r.$$

By induction,

$$f(x_k) = f([k-1]\Delta x_k) = [(k-1)h]^r = h^r(k-1)^r,$$

So,

$$f(x_k)\Delta x_k = h^{r+1}(k-1)^r$$
$$\sum_{k=1}^{n} f(x_k)\Delta x_k = h^{r+1}\sum_{k=1}^{n}(k-1)^r.$$

For example, take r=2.

$$\sum_{k=1}^{n} f(x_k)\Delta x_k = h^3\sum_{k=1}^{n}(k-1)^2$$
$$= h^3\left(\sum_{k=1}^{n}k^2 - 2\sum_{k=1}^{n}k + \sum_{k=1}^{n}1\right)$$
$$= h^3\left[\frac{n(n+1)(2n+1)}{6} - 2\frac{n(n+1)}{2} + n\right],$$

So,

$$I \equiv \lim_{n\to\infty}\sum_{k=1}^{n} f(x_k^*)\Delta x_k = \lim_{n\to\infty}\sum_{k=1}^{n} f(x_k)\Delta x_k$$
$$= \lim_{n\to\infty} h^3\left[\frac{n(n+1)(2n+1)}{6} - 2\frac{n(n+1)}{2} + n\right],$$
$$= a^3\lim_{n\to\infty}\left[\frac{n(n+1)(2n+1)}{6n^3} - \frac{n(n+1)}{n^3} + \frac{n}{n^3}\right],$$
$$= \frac{1}{3}a^3.$$

Riemann integrals can be computed only for proper integrals.

References

- Integration, maths: byjus.com, Retrieved 08, August 2020
- Formulas-of-integration, math-results-and-formulas: emathzone.com, Retrieved 16, May 2020
- Latex-source-and-diagrams, Engineering-maths-first-aid-kit: mathcentre.ac.uk, Retrieved 07, March 2020
- RiemannIntegral: wolfram.com, Retrieved 19, July 2020

5

Techniques of Integration

There are various methods which are used for integration such as disc integration, shell integration, integration by substitution, trigonometric substitution, order of integration and integration by reduction. The topics elaborated in this chapter will help in gaining a better perspective about these methods of integration.

Methods of Contour Integration

In the mathematical field of complex analysis, contour integration is a method of evaluating certain integrals along paths in the complex plane.

Contour integration is closely related to the calculus of residues, a method of complex analysis.

One use for contour integrals is the evaluation of integrals along the real line that are not readily found by using only real variable methods.

Contour integration methods include:

- direct integration of a complex-valued function along a curve in the complex plane (a *contour*).
- application of the Cauchy integral formula.
- application of the residue theorem.

One method can be used, or a combination of these methods, or various limiting processes, for the purpose of finding these integrals or sums.

Curves in the Complex Plane

In complex analysis a contour is a type of curve in the complex plane. In contour integration, contours provide a precise definition of the curves on which an integral may be suitably defined. A curve in the complex plane is defined as a continuous function from a closed interval of the real line to the complex plane: $z : [a, b] \to \mathbf{C}$.

This definition of a curve coincides with the intuitive notion of a curve, but includes a parametrization by a continuous function from a closed interval. This more precise definition allows us to consider what properties a curve must have for it to be useful for integration. In the following subsections we narrow down the set of curves that we can integrate to only include ones that can be built up out of a finite number of continuous curves that can be given a direction. Moreover, we will restrict the "pieces" from crossing over themselves, and we require that each piece have a finite (non-vanishing) continuous derivative. These requirements correspond to requiring that we consider only curves that can be traced, such as by a pen, in a sequence of even, steady strokes, which only stop to start a new piece of the curve, all without picking up the pen.

Directed Smooth Curves

Contours are often defined in terms of directed smooth curves. These provide a precise definition of a "piece" of a smooth curve, of which a contour is made.

A smooth curve is a curve $z : [a, b] \to C$ with a non-vanishing, continuous derivative such that each point is traversed only once (z is one-to-one), with the possible exception of a curve such that the endpoints match ($z(a) = z(b)$). In the case where the endpoints match the curve is called closed, and the function is required to be one-to-one everywhere else and the derivative must be continuous at the identified point ($z'(a) = z'(b)$). A smooth curve that is not closed is often referred to as a smooth arc.

The parametrization of a curve provides a natural ordering of points on the curve: $z(x)$ comes before $z(y)$ if $x < y$. This leads to the notion of a directed smooth curve. It is most useful to consider curves independent of the specific parametrization. This can be done by considering equivalence classes of smooth curves with the same direction. A directed smooth curve can then be defined as an ordered set of points in the complex plane that is the image of some smooth curve in their natural order (according to the parametrization). Note that not all orderings of the points are the natural ordering of a smooth curve. In fact, a given smooth curve has only two such orderings. Also, a single closed curve can have any point as its endpoint, while a smooth arc has only two choices for its endpoints.

Contours

Contours are the class of curves on which we define contour integration. A contour is a directed curve which is made up of a finite sequence of directed smooth curves whose endpoints are matched to give a single direction. This requires that the sequence of curves $\gamma_1,\ldots,\gamma_n$ be such that the terminal point of γ_i coincides with the initial point of γ_{i+1}, $\forall\, i, 1 \le i < n$. This includes all directed smooth curves. Also, a single point in the complex plane is considered a contour. The symbol + is often used to denote the piecing of curves together to form a new curve. Thus we could write a contour Γ that is made up of n contours as:

$$\Gamma = \gamma_1 + \gamma_2 + \cdots + \gamma_n.$$

Contour Integrals

The contour integral of a complex function $f : C \to \mathbf{C}$ is a generalization of the integral for real-valued functions. For continuous functions in the complex plane, the contour integral can be defined in analogy to the line integral by first defining the integral along a directed smooth curve in terms of an integral over a real valued parameter. A more general definition can be given in terms of partitions of the contour in analogy with the partition of an interval and the Riemann integral. In both cases the integral over a contour is defined as the sum of the integrals over the directed smooth curves that make up the contour.

For Continuous Functions

To define the contour integral in this way one must first consider the integral, over a real variable, of a complex-valued function. Let $f: \mathbb{R} \to \mathbb{C}$ be a complex-valued function of a real variable, t. The real and imaginary parts of f are often denoted as $u(t)$ and $v(t)$, respectively, so that:

$$f(t) = u(t) + iv(t).$$

Then the integral of the complex-valued function f over the interval $[a, b]$ is given by:

$$\int_a^b f(t)dt = \int_a^b \big(u(t) + iv(t)\big)dt$$
$$= \int_a^b u(t)dt + i\int_a^b v(t)dt.$$

Let $f: \mathbb{C} \to \mathbb{C}$ be a continuous function on the directed smooth curve γ. Let $z: \mathbb{R} \to \mathbb{C}$ be any parametrization of γ that is consistent with its order (direction). Then the integral along γ is denoted:

$$\int_\gamma f(z)dz$$

and is given by:

$$\int_\gamma f(z)dz = \int_a^b f\big(z(t)\big)z'(t)dt.$$

This definition is well defined. That is, the result is independent of the parametrization chosen. In the case where the real integral on the right side does not exist the integral along γ is said not to exist.

As a Generalization of the Riemann Integral

The generalization of the Riemann integral to functions of a complex variable is done in complete analogy to its definition for functions from the real numbers. The partition of a directed smooth curve γ is defined as a finite, ordered set of points on γ. The integral over the curve is the limit of finite sums of function values, taken at the points on the partition, in the limit that the maximum distance between any two successive points on the partition (in the two-dimensional complex plane), also known as the mesh, goes to zero.

Direct Methods

Direct methods involve the calculation of the integral by means of methods similar to those in calculating line integrals in several-variable calculus. This means that we use the following method:

- Parametrizing the contour.

  The contour is parametrized by a differentiable complex-valued function of real variables, or the contour is broken up into pieces and parametrized separately.

- Substitution of the parametrization into the integrand.

Substituting the parametrization into the integrand transforms the integral into an integral of one real variable.

- Direct evaluation.

  The integral is evaluated in a method akin to a real-variable integral.

Example

A fundamental result in complex analysis is that the contour integral of $1/z$ is $2\pi i$, where the path of the contour is taken to be the unit circle traversed counterclockwise (or any positively oriented Jordan curve about 0). In the case of the unit circle there is a direct method to evaluate the integral:

$$\oint_C \frac{1}{z}dz.$$

In evaluating this integral, use the unit circle $|z| = 1$ as contour, parametrized by $z(t) = e^{it}$, with $t \in [0, 2\pi]$, then $dz/dt = ie^{it}$ and,

$$\oint_C \frac{1}{z}dz = \int_0^{2\pi} \frac{1}{e^{it}} ie^{it}\,dt = i\int_0^{2\pi} 1dt = \left[\,t\,\right]_0^{2\pi} i = (2\pi - 0)i = 2\pi i.$$

which is the value of the integral.

Applications of Integral Theorems

Applications of integral theorems are also often used to evaluate the contour integral along a contour, which means that the real-valued integral is calculated simultaneously along with calculating the contour integral.

Integral theorems such as the Cauchy integral formula or residue theorem are generally used in the following method:

- A specific contour is chosen.

  The contour is chosen so that the contour follows the part of the complex plane that describes the real-valued integral, and also encloses singularities of the integrand so application of the Cauchy integral formula or residue theorem is possible.

- Application of Cauchy's integral theorem.

  The integral is reduced to only an integration around a small circle about each pole.

- Application of the Cauchy integral formula or residue theorem.

  Application of these integral formula gives us a value for the integral around the whole of the contour.

- Division of the contour into a contour along the real part and imaginary part.

  The whole of the contour can be divided into the contour that follows the part of the complex plane that describes the real-valued integral as chosen before (call it R), and the integral that crosses the complex plane (call it *I*). The integral over the whole of the contour is the sum of the integral over each of these contours.

- Demonstration that the integral that crosses the complex plane plays no part in the sum.

  If the integral I can be shown to be zero, or if the real-valued integral that is sought is improper, then if we demonstrate that the integral I as described above tends to 0, the integral along R will tend to the integral around the contour $R + I$.

- Conclusion.

  If we can show the above step, then we can directly calculate R, the real-valued integral.

Example 1

Consider the integral:

$$\int_{-\infty}^{\infty} \frac{1}{\left(x^2+1\right)^2} dx,$$

To evaluate this integral, we look at the complex-valued function:

$$f(z) = \frac{1}{\left(z^2+1\right)^2}$$

which has singularities at i and $-i$. We choose a contour that will enclose the real-valued integral, here a semicircle with boundary diameter on the real line (going from, say, $-a$ to a) will be convenient. Call this contour C.

There are two ways of proceeding, using the Cauchy integral formula or by the method of residues:

Using the Cauchy Integral Formula

Note that:

$$\oint_C f(z)dz = \int_{-a}^{a} f(z)dz + \int_{\text{Arc}} f(z)dz$$

thus,

$$\int_{-a}^{a} f(z)dz = \oint_C f(z)dz - \int_{\text{Arc}} f(z)dz$$

Furthermore observe that:

$$f(z) = \frac{1}{\left(z^2+1\right)^2} = \frac{1}{(z+i)^2(z-i)^2}.$$

Since the only singularity in the contour is the one at i, then we can write,

$$f(z) = \frac{\frac{1}{(z+i)^2}}{(z-i)^2},$$

which puts the function in the form for direct application of the formula. Then, by using Cauchy's integral formula,

$$\oint_C f(z)dz = \oint_C \frac{\frac{1}{(z+i)^2}}{(z-i)^2}dz = 2\pi i \frac{d}{dz}\left(\frac{1}{(z+i)^2}\right)\Big|_{z=i} = 2\pi i\left(\frac{-2}{(z+i)^3}\right)\Big|_{z=i} = \frac{\pi}{2}$$

We take the first derivative, in the above steps, because the pole is a second-order pole. That is, $(z - i)$ is taken to the second power, so we employ the first derivative of $f(z)$. If it were $(z - i)$ taken to the third power, we would use the second derivative and divide by 2!, etc. The case of $(z - i)$ to the first power corresponds to a zero order derivative—just $f(z)$ itself.

We need to show that the integral over the arc of the semicircle tends to zero as $a \to \infty$, using the estimation lemma,

$$\left|\int_{\text{Arc}} f(z)dz\right| \le ML$$

where M is an upper bound on $|f(z)|$ along the arc and L the length of the arc. Now,

$$\left|\int_{\text{Arc}} f(z)dz\right| \le \frac{a\pi}{\left(a^2-1\right)^2} \to 0 \text{ as } a \to \infty.$$

So,

$$\int_{-\infty}^{\infty} \frac{1}{\left(x^2+1\right)^2}dx = \int_{-\infty}^{\infty} f(z)dz = \lim_{a\to+\infty} \int_{-a}^{a} f(z)dz = \frac{\pi}{2}.$$

Using the Method of Residues

Consider the Laurent series of $f(z)$ about i, the only singularity we need to consider. We then have,

$$f(z) = \frac{-1}{4(z-i)^2} + \frac{-i}{4(z-i)} + \frac{3}{16} + \frac{i}{8}(z-i) + \frac{-5}{64}(z-i)^2 + \cdots$$

It is clear by inspection that the residue is $-i/4$ (to see this, imagine that the above equation were multiplied by $z - i$, then both sides integrated via the Cauchy integral formula—only the second term would integrate to a non-zero quantity), so, by the residue theorem, we have,

$$\oint_C f(z)dz = \oint_C \frac{1}{\left(z^2+1\right)^2}dz = 2\pi i \operatorname{Res}_{z=i} f = 2\pi i\left(-\frac{i}{4}\right) = \frac{\pi}{2} \quad \square$$

Thus we get the same result as before.

Contour Note

As an aside, a question can arise whether we do not take the semicircle to include the *other* singularity, enclosing $-i$. To have the integral along the real axis moving in the correct direction, the contour must travel clockwise, i.e., in a negative direction, reversing the sign of the integral overall.

This does not affect the use of the method of residues by series.

Example 2 – Cauchy Distribution

The integral,

$$\int_{-\infty}^{\infty} \frac{e^{itx}}{x^2+1} dx$$

(which arises in probability theory as a scalar multiple of the characteristic function of the Cauchy distribution) resists the techniques of elementary calculus. We will evaluate it by expressing it as a limit of contour integrals along the contour C that goes along the real line from $-a$ to a and then counterclockwise along a semicircle centered at 0 from a to $-a$. Take a to be greater than 1, so that the imaginary unit i is enclosed within the curve. The contour integral is:

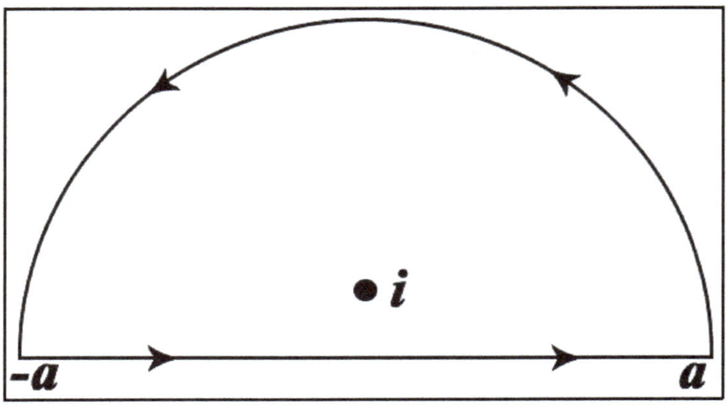

$$\int_C \frac{e^{itz}}{z^2+1} dz.$$

Since e^{itz} is an entire function (having no singularities at any point in the complex plane), this function has singularities only where the denominator $z^2 + 1$ is zero. Since $z^2 + 1 = (z + i)(z - i)$, that happens only where $z = i$ or $z = -i$. Only one of those points is in the region bounded by this contour. The residue of $f(z)$ at $z = i$ is:

$$\lim_{z\to i}(z-i)f(z) = \lim_{z\to i}(z-i)\frac{e^{itz}}{z^2+1} = \lim_{z\to i}(z-i)\frac{e^{itz}}{(z-i)(z+i)} = \lim_{z\to i}\frac{e^{itz}}{z+i} = \frac{e^{-t}}{2i}.$$

According to the residue theorem, then, we have,

$$\int_C f(z)dz = 2\pi i \operatorname{Res}_{z=i} f(z) = 2\pi i \frac{e^{-t}}{2i} = \pi e^{-t}.$$

The contour C may be split into a "straight" part and a curved arc, so that,

$$\int_{\text{straight}} + \int_{\text{arc}} = \pi e^{-t},$$

and thus,

$$\int_{-a}^{a} = \pi e^{-t} - \int_{\text{arc}}.$$

It can be shown that if t > 0 then,

$$\int_{\text{arc}} \frac{e^{itz}}{z^2+1}\,dz \to 0 \text{ as } a \to \infty.$$

Therefore if t > 0 then,

$$\int_{-\infty}^{\infty} \frac{e^{itx}}{x^2+1}\,dx = \pi e^{-t}.$$

A similar argument with an arc that winds around $-i$ rather than i shows that if t < 0 then,

$$\int_{-\infty}^{\infty} \frac{e^{itx}}{x^2+1}\,dx = \pi e^{t},$$

and finally we have this:

$$\int_{-\infty}^{\infty} \frac{e^{itx}}{x^2+1}\,dx = \pi e^{-|t|}. \quad \square$$

(If $t = 0$ then the integral yields immediately to real-valued calculus methods and its value is π.)

Example 3 – Trigonometric Integrals

Certain substitutions can be made to integrals involving trigonometric functions, so the integral is transformed into a rational function of a complex variable and then the above methods can be used in order to evaluate the integral.

As an example, consider,

$$\int_{-\pi}^{\pi} \frac{1}{1+3(\cos t)^2}\,dt.$$

We seek to make a substitution of $z = e^{it}$. Now, recall,

$$\cos t = \frac{1}{2}\left(e^{it} + e^{-it}\right) = \frac{1}{2}\left(z + \frac{1}{z}\right)$$

and

$$\frac{dz}{dt} = iz,\ dt = \frac{dz}{iz}.$$

Taking C to be the unit circle, we substitute to get:

$$\oint_C \frac{1}{1+3\left(\frac{1}{2}\left(z+\frac{1}{z}\right)\right)^2}\frac{dz}{iz} = \oint_C \frac{1}{1+\frac{3}{4}\left(z+\frac{1}{z}\right)^2}\frac{1}{iz}dz$$

$$= \oint_C \frac{-i}{z+\frac{3}{4}z\left(z+\frac{1}{z}\right)^2}dz$$

$$= -i\oint_C \frac{1}{z+\frac{3}{4}z\left(z^2+2+\frac{1}{z^2}\right)}dz$$

$$= -i\oint_C \frac{1}{z+\frac{3}{4}\left(z^3+2z+\frac{1}{z}\right)}dz$$

$$= -i\oint_C \frac{1}{\frac{3}{4}z^3+\frac{5}{2}z+\frac{3}{4z}}dz$$

$$= -i\oint_C \frac{4}{3z^3+10z+\frac{3}{z}}dz$$

$$= -4i\oint_C \frac{1}{3z^3+10z+\frac{3}{z}}dz \quad , \quad = -4i\oint_C \frac{z}{3z^4+10z^2+3}dz$$

$$= -4i\oint_C \frac{z}{3\left(z+\sqrt{3}i\right)\left(z-\sqrt{3}i\right)\left(z+\frac{i}{\sqrt{3}}\right)\left(z-\frac{i}{\sqrt{3}}\right)}dz$$

$$= -\frac{4}{3}i\oint_C \frac{z}{\left(z+\sqrt{3}i\right)\left(z-\sqrt{3}i\right)\left(z+\frac{i}{\sqrt{3}}\right)\left(z-\frac{i}{\sqrt{3}}\right)}dz.$$

$$= \frac{8\pi}{3}\left[\frac{3}{16}+\frac{3}{16}\right] \quad , \quad = \pi.$$

The singularities to be considered are at $3^{-1/2}i$, $-3^{-1/2}i$. Let C_1 be a small circle about $3^{-1/2}i$, and C_2 be a small circle about $-3^{-1/2}i$. Then we arrive at the following:

Example 3a – Trigonometric Integrals, the General Procedure

The above method may be applied to all integrals of the type,

$$\int_0^{2\pi} \frac{P\left(\sin(t),\sin(2t),\ldots,\cos(t),\cos(2t),\ldots\right)}{Q\left(\sin(t),\sin(2t),\ldots,\cos(t),\cos(2t),\ldots\right)}dt$$

where P and Q are polynomials, i.e. a rational function in trigonometric terms is being integrated. Note that the bounds of integration may as well be π and $-\pi$, as in the previous example, or any other pair of endpoints 2π apart.

The trick is to use the substitution $z = e^{it}$ where $dz = ie^{it}\, dt$ and hence,

$$\frac{1}{iz}dz = dt.$$

This substitution maps the interval [0, 2π] to the unit circle. Furthermore,

$$\sin(kt) = \frac{e^{ikt} - e^{-ikt}}{2i} = \frac{z^k - z^{-k}}{2i}$$

and

$$\cos(kt) = \frac{e^{ikt} + e^{-ikt}}{2} = \frac{z^k + z^{-k}}{2}$$

so that a rational function $f(z)$ in z results from the substitution, and the integral becomes,

$$\oint_{|z|=1} f(z)\frac{1}{iz}dz$$

which is in turn computed by summing the residues of $f(z)1/iz$ inside the unit circle.

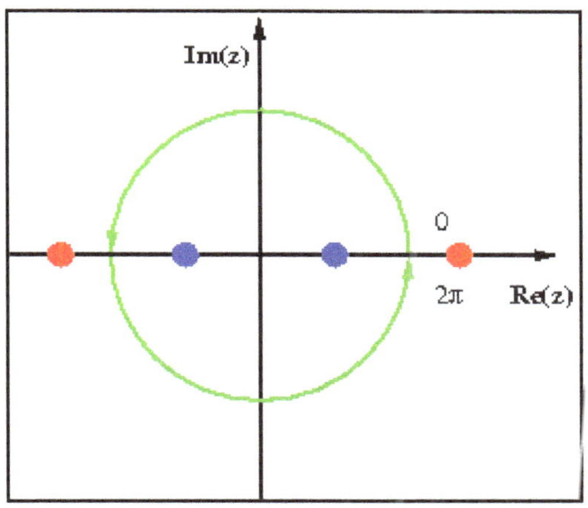

$$I = \int_0^{\frac{\pi}{2}} \frac{1}{1+(\sin t)^2}dt,$$

which we now compute. The first step is to recognize that,

$$I = \frac{1}{4}\int_0^{2\pi} \frac{1}{1+(\sin t)^2}dt.$$

The substitution yields,

$$\frac{1}{4}\oint_{|z|=1} \frac{4iz}{z^4 - 6z^2 + 1}dz = \oint_{|z|=1} \frac{iz}{z^4 - 6z^2 + 1}dz.$$

The poles of this function are at $1 \pm \sqrt{2}$ and $-1 \pm \sqrt{2}$. Of these, $1 + \sqrt{2}$ and $-1 - \sqrt{2}$ are outside the unit circle (shown in red, not to scale), whereas $1 - \sqrt{2}$ and $-1 + \sqrt{2}$ are inside the unit circle (shown in blue). The corresponding residues are both equal to $-i\sqrt{2}/16$, so that the value of the integral is:

$$I = 2\pi i \, 2\left(-\frac{\sqrt{2}}{16}i\right) = \pi\frac{\sqrt{2}}{4}.$$

Example 4 – Branch Cuts

Consider the real integral,

$$\int_0^\infty \frac{\sqrt{x}}{x^2+6x+8}dx.$$

We can begin by formulating the complex integral,

$$\int_C \frac{\sqrt{z}}{z^2+6z+8}dz = I.$$

We can use the Cauchy integral formula or residue theorem again to obtain the relevant residues. However, the important thing to note is that $z^{1/2} = e^{\frac{1}{2}\operatorname{Log} z}$, so $z^{1/2}$ has a branch cut. This affects our choice of the contour *C*. Normally the logarithm branch cut is defined as the negative real axis, however, this makes the calculation of the integral slightly more complicated, so we define it to be the positive real axis.

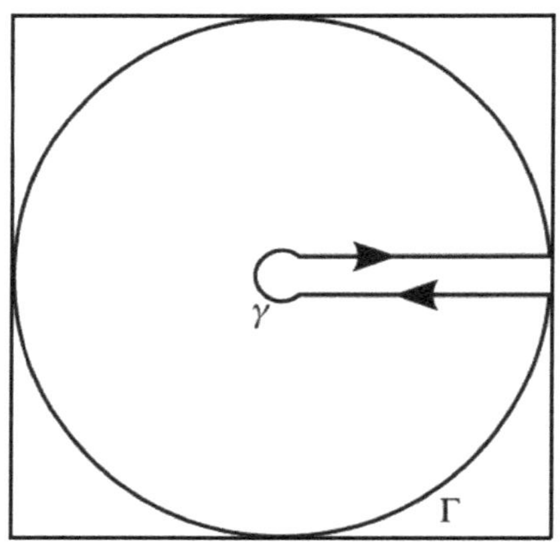

Then, we use the so-called *keyhole contour*, which consists of a small circle about the origin of radius ε say, extending to a line segment parallel and close to the positive real axis but not touching it, to an almost full circle, returning to a line segment parallel, close, and below the positive real axis in the negative sense, returning to the small circle in the middle.

Note that $z = -2$ and $z = -4$ are inside the big circle. These are the two remaining poles, derivable by factoring the denominator of the integrand. The branch point at $z = 0$ was avoided by detouring around the origin.

Let γ be the small circle of radius ε, Γ the larger, with radius R, then,

$$\int_C = \int_\varepsilon^R + \int_\Gamma + \int_R^\varepsilon + \int_\gamma.$$

It can be shown that the integrals over Γ and γ both tend to zero as $\varepsilon \to 0$ and $R \to \infty$, by an estimation argument above, that leaves two terms. Now since $z^{1/2} = e^{1/2 \operatorname{Log} z}$, on the contour outside the branch cut, we have gained 2π in argument along γ. (By Euler's identity, $e^{i\pi}$ represents the unit vector, which therefore has π as its log. This π is what is meant by the argument of z. The coefficient of 1/2 forces us to use 2π.) So,

$$\begin{aligned}
\int_R^\varepsilon \frac{\sqrt{z}}{z^2+6z+8}dz &= \int_R^\varepsilon \frac{e^{\frac{1}{2}\operatorname{Log} z}}{z^2+6z+8}dz \\
&= \int_R^\varepsilon \frac{e^{\frac{1}{2}(\log|z|+i\arg z)}}{z^2+6z+8}dz \\
&= \int_R^\varepsilon \frac{e^{\frac{1}{2}\log|z|}e^{\frac{1}{2}(2\pi i)}}{z^2+6z+8}dz \\
&= \int_R^\varepsilon \frac{e^{\frac{1}{2}\log|z|}e^{\pi i}}{z^2+6z+8}dz \\
&= \int_R^\varepsilon \frac{-\sqrt{z}}{z^2+6z+8}dz \\
&= \int_\varepsilon^R \frac{\sqrt{z}}{z^2+6z+8}dz.
\end{aligned}$$

Therefore:

$$\int_C \frac{\sqrt{z}}{z^2+6z+8}dz = 2\int_0^\infty \frac{\sqrt{x}}{x^2+6x+8}dx.$$

By using the residue theorem or the Cauchy integral formula (first employing the partial fractions method to derive a sum of two simple contour integrals) one obtains,

$$\pi i\left(\frac{i}{\sqrt{2}}-i\right) = \int_0^\infty \frac{\sqrt{x}}{x^2+6x+8}dx = \pi\left(1-\frac{1}{\sqrt{2}}\right).$$

Example 5 – The Square of the Logarithm

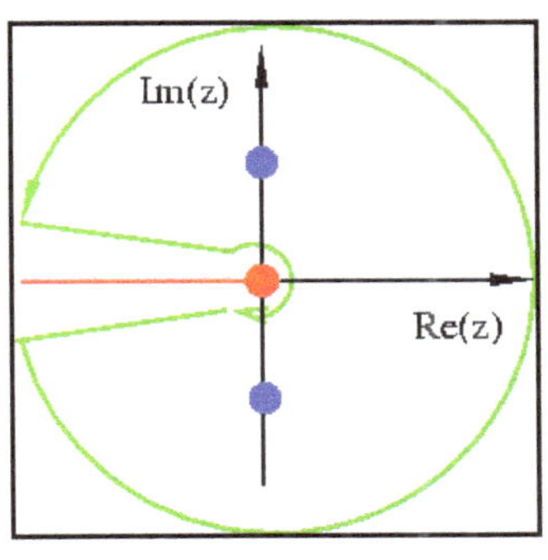

This section treats a type of integral of which is an example.

$$\int_0^\infty \frac{\log x}{\left(1+x^2\right)^2}dx$$

To calculate this integral, one uses the function:

$$f(z) = \left(\frac{\log z}{1+z^2}\right)^2$$

and the branch of the logarithm corresponding to $-\pi < \arg z \leq \pi$.

We will calculate the integral of $f(z)$ along the keyhole contour shown at right. As it turns out this integral is a multiple of the initial integral that we wish to calculate and by the Cauchy residue theorem we have,

$$\begin{aligned}\left(\int_R+\int_M+\int_N+\int_r\right)f(z)dz &= 2\pi i\left(\text{Res}_{z=i} f(z)+\text{Res}_{z=-i} f(z)\right)\\ &= 2\pi i\left(-\frac{\pi}{4}+\frac{1}{16}i\pi^2-\frac{\pi}{4}-\frac{1}{16}i\pi^2\right)\\ &= -i\pi^2.\end{aligned}$$

Let R be the radius of the large circle, and r the radius of the small one. We will denote the upper line by M, and the lower line by N. As before we take the limit when $R \to \infty$ and $r \to 0$. The contributions from the two circles vanish. For example, one has the following upper bound with the ML lemma:

$$\left|\int_R f(z)dz\right| \leq 2\pi R\frac{(\log R)^2+\pi^2}{\left(R^2-1\right)^2} \to 0.$$

In order to compute the contributions of M and N we set $z = -x + i\varepsilon$ on M and $z = -x - i\varepsilon$ on N, with $0 < x < \infty$:

$$\begin{aligned}-i\pi^2 &= \left(\int_R+\int_M+\int_N+\int_r\right)f(z)dz\\ &= \left(\int_M+\int_N\right)f(z)dz \int_R,\int_r \text{ vanish}\\ &= -\int_\infty^0\left(\frac{\log(-x+i\varepsilon)}{1+(-x+i\varepsilon)^2}\right)^2dx-\int_0^\infty\left(\frac{\log(-x-i\varepsilon)}{1+(-x-i\varepsilon)^2}\right)^2dx\\ &= \int_0^\infty\left(\frac{\log(-x+i\varepsilon)}{1+(-x+i\varepsilon)^2}\right)^2dx-\int_0^\infty\left(\frac{\log(-x-i\varepsilon)}{1+(-x-i\varepsilon)^2}\right)^2dx\\ &= \int_0^\infty\left(\frac{\log x+i\pi}{1+x^2}\right)^2dx-\int_0^\infty\left(\frac{\log x-i\pi}{1+x^2}\right)^2dx\varepsilon \to 0\\ &= \int_0^\infty\frac{(\log x+i\pi)^2-(\log x-i\pi)^2}{\left(1+x^2\right)^2}dx\\ &= \int_0^\infty\frac{4\pi i\log x}{\left(1+x^2\right)^2}dx\\ &= 4\pi i\int_0^\infty\frac{\log x}{\left(1+x^2\right)^2}dx\end{aligned}$$

which gives,

$$\int_0^{\infty} \frac{\log x}{\left(1+x^2\right)^2} dx = -\frac{\pi}{4}.$$

Example 6 – Logarithms and the Residue at Infinity

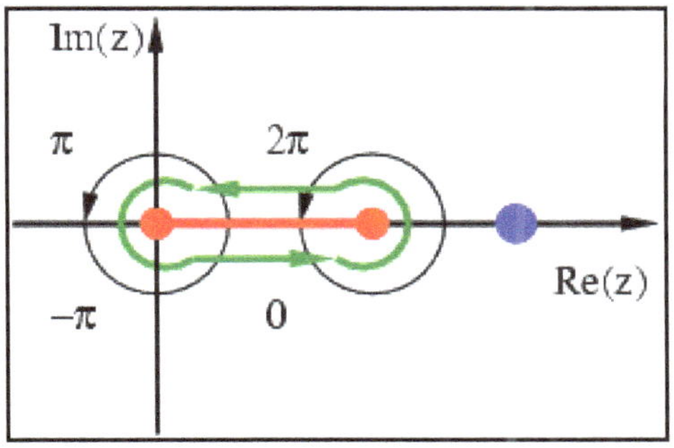

We seek to evaluate.

$$I = \int_0^3 \frac{x^{\frac{3}{4}}(3-x)^{\frac{1}{4}}}{5-x} dx.$$

This requires a close study of:

$$f(z) = z^{\frac{3}{4}}(3-z)^{\frac{1}{4}}.$$

We will construct $f(z)$ so that it has a branch cut on [0, 3], shown in red in the diagram. To do this, we choose two branches of the logarithm, setting,

$$z^{\frac{3}{4}} = \exp\left(\frac{3}{4}\log z\right) \quad \text{where } -\pi \le \arg z < \pi$$

and

$$(3-z)^{\frac{1}{4}} = \exp\left(\frac{1}{4}\log(3-z)\right) \quad \text{where } 0 \le \arg(3-z) < 2\pi.$$

The cut of $z^{3/4}$ is therefore $(-\infty, 0]$ and the cut of $(3-z)^{1/4}$ is $(-\infty, 3]$. It is easy to see that the cut of the product of the two, i.e. $f(z)$, is [0, 3], because $f(z)$ is actually continuous across $(-\infty, 0)$. This is because when $z = -r < 0$ and we approach the cut from above, $f(z)$ has the value

$$r^{\frac{3}{4}} e^{\frac{3}{4}\pi i} (3+r)^{\frac{1}{4}} e^{\frac{2}{4}\pi i} = r^{\frac{3}{4}} (3+r)^{\frac{1}{4}} e^{\frac{5}{4}\pi i}.$$

When we approach from below, $f(z)$ has the value,

$$r^{\frac{3}{4}} e^{-\frac{3}{4}\pi i} (3+r)^{\frac{1}{4}} e^{\frac{0}{4}\pi i} = r^{\frac{3}{4}} (3+r)^{\frac{1}{4}} e^{-\frac{3}{4}\pi i}.$$

But,

$$e^{-\frac{3}{4}\pi i} = e^{\frac{5}{4}\pi i},$$

so that we have continuity across the cut. This is illustrated in the diagram, where the two black oriented circles are labelled with the corresponding value of the argument of the logarithm used in $z^{3/4}$ and $(3 - z)^{1/4}$.

We will use the contour shown in green in the diagram. To do this we must compute the value of $f(z)$ along the line segments just above and just below the cut.

Let $z = r$ (in the limit, i.e. as the two green circles shrink to radius zero), where $0 \le r \le 3$. Along the upper segment, we find that $f(z)$ has the value,

$$r^{\frac{3}{4}}e^{\frac{0}{4}\pi i}(3-r)^{\frac{1}{4}}e^{\frac{2}{4}\pi i} = ir^{\frac{3}{4}}(3-r)^{\frac{1}{4}}$$

and along the lower segment,

$$r^{\frac{3}{4}}e^{\frac{0}{4}\pi i}(3-r)^{\frac{1}{4}}e^{\frac{0}{4}\pi i} = r^{\frac{3}{4}}(3-r)^{\frac{1}{4}}.$$

It follows that the integral of $f(z)/5 - z$ along the upper segment is $-iI$ in the limit, and along the lower segment, I.

If we can show that the integrals along the two green circles vanish in the limit, then we also have the value of I, by the Cauchy residue theorem. Let the radius of the green circles be ρ, where $\rho < 0.001$ and $\rho \to 0$, and apply the ML inequality. For the circle C_L on the left, we find:

$$\left|\int_{C_L} \frac{f(z)}{5-z}dz\right| \le 2\pi\rho\frac{\rho^{\frac{3}{4}}3.001^{\frac{1}{4}}}{4.999} \in \mathcal{O}\left(\rho^{\frac{7}{4}}\right) \to 0.$$

Similarly, for the circle C_R on the right, we have,

$$\left|\int_{C_R} \frac{f(z)}{5-z}dz\right| \le 2\pi\rho\frac{3.001^{\frac{3}{4}}\rho^{\frac{1}{4}}}{1.999} \in \mathcal{O}\left(\rho^{\frac{5}{4}}\right) \to 0.$$

Now using the Cauchy residue theorem, we have,

$$(-i+1)I = -2\pi i\left(\operatorname{Res}_{z=5}\frac{f(z)}{5-z} + \operatorname{Res}_{z=\infty}\frac{f(z)}{5-z}\right).$$

where the minus sign is due to the clockwise direction around the residues. Using the branch of the logarithm from before, clearly,

$$\operatorname{Res}_{z=5}\frac{f(z)}{5-z} = -5^{\frac{3}{4}}e^{\frac{1}{4}\log(-2)}.$$

The pole is shown in blue in the diagram. The value simplifies to,

$$-5^{\frac{3}{4}}e^{\frac{1}{4}(\log 2+\pi i)} = -e^{\frac{1}{4}\pi i}5^{\frac{3}{4}}2^{\frac{1}{4}}.$$

We use the following formula for the residue at infinity:

$$\mathrm{Res}_{z=\infty}\, h(z) = \mathrm{Res}_{z=0}\left(-\frac{1}{z^2}h\left(\frac{1}{z}\right)\right).$$

Substituting, we find:

$$\frac{1}{5-\frac{1}{z}} = -z\left(1+5z+5^2z^2+5^3z^3+\cdots\right)$$

and

$$\left(\frac{1}{z^3}\left(3-\frac{1}{z}\right)\right)^{\frac{1}{4}} = \frac{1}{z}(3z-1)^{\frac{1}{4}} = \frac{1}{z}e^{\frac{1}{4}\pi i}(1-3z)^{\frac{1}{4}},$$

where we have used the fact that $-1 = e^{\pi i}$ for the second branch of the logarithm. Next we apply the binomial expansion, obtaining:

$$\frac{1}{z}e^{\frac{1}{4}\pi i}\left(1-\binom{\frac{1}{4}}{1}3z+\binom{\frac{1}{4}}{2}3^2z^2-\binom{\frac{1}{4}}{3}3^3z^3+\cdots\right).$$

The conclusion is that,

$$\mathrm{Res}_{z=\infty}\frac{f(z)}{5-z} = e^{\frac{1}{4}\pi i}\left(5-\frac{3}{4}\right) = e^{\frac{1}{4}\pi i}\frac{17}{4}.$$

Finally, it follows that the value of I is:

$$I = 2\pi i\frac{e^{\frac{1}{4}\pi i}}{-1+i}\left(\frac{17}{4}-5^{\frac{3}{4}}2^{\frac{1}{4}}\right) = 2\pi 2^{-\frac{1}{2}}\left(\frac{17}{4}-5^{\frac{3}{4}}2^{\frac{1}{4}}\right)$$

which yields,

$$I = \frac{\pi}{2\sqrt{2}}\left(17-5^{\frac{3}{4}}2^{\frac{9}{4}}\right) = \frac{\pi}{2\sqrt{2}}\left(17-40^{\frac{3}{4}}\right).$$

Integral Representation

An integral representation of a function is an expression of the function involving a contour integral. Various integral representations are known for many special functions. Integral representations can be important for theoretical reasons, e.g. giving analytic continuation or functional equations, or sometimes for numerical evaluations.

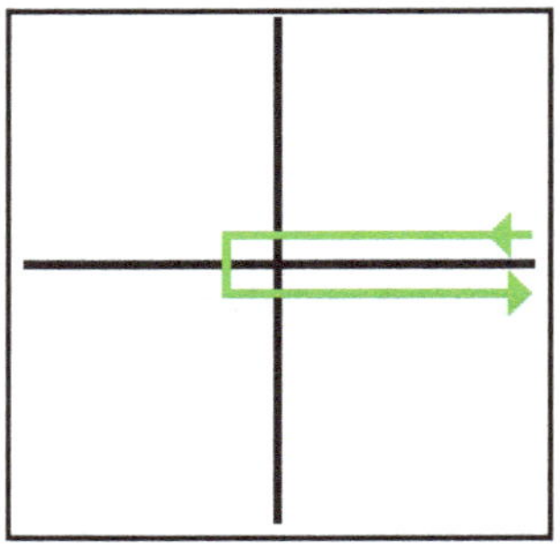

Hankel's contour.

For example, the original definition of the Riemann zeta function $\zeta(s)$ via a Dirichlet series,

$$\zeta(s) = \sum_{n=1}^{\infty} \frac{1}{n^s},$$

is valid only for Re(s) > 1. But,

$$\zeta(s) = -\frac{\Gamma(1-s)}{2\pi i} \int_H \frac{(-t)^{s-1}}{e^t - 1} dt,$$

where the integration is done over the Hankel contour, H, is valid for all complex $s \neq 1$.

Disc Integration

Disc integration, also known in integral calculus as the disc method, is a means of calculating the volume of a solid of revolution of a solid-state material when integrating along the axis of revolution. This method models the resulting three-dimensional shape as a stack of an infinite number of discs of varying radius and infinitesimal thickness. It is also possible to use the same principles with rings instead of discs (the "washer method") to obtain hollow solids of revolutions.

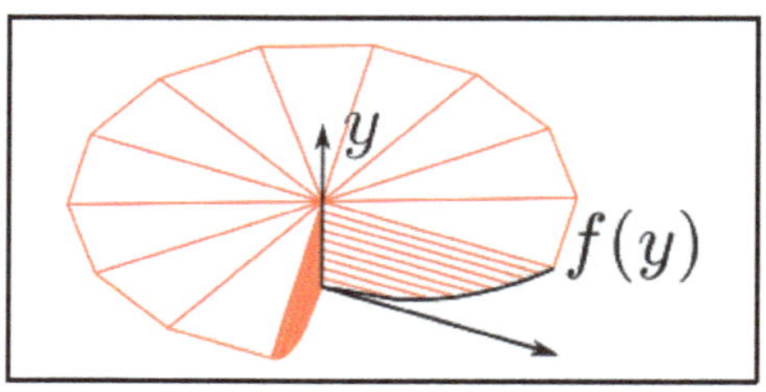

Definition

Function of *x*

If the function to be revolved is a function of x, the following integral represents the volume of the solid of revolution.

$$\pi\int_a^b R(x)^2\,dx$$

where $R(x)$ is the distance between the function and the axis of rotation. This works only if the axis of rotation is horizontal (example: $y = 3$ or some other constant).

Function of *y*

If the function to be revolved is a function of y, the following integral will obtain the volume of the solid of revolution:

$$\pi\int_c^d R(y)^2\,dy$$

where $R(y)$ is the distance between the function and the axis of rotation. This works only if the axis of rotation is vertical (example: $x = 4$ or some other constant).

Washer Method

To obtain a hollow solid of revolution (the "washer method"), the procedure would be to take the volume of the inner solid of revolution and subtract it from the volume of the outer solid of revolution. This can be calculated in a single integral similar to the following:

$$\pi\int_a^b \left(R_O(x)^2 - R_I(x)^2\right)dx$$

where $R_O(x)$ is the function that is farthest from the axis of rotation and $R_I(x)$ is the function that is closest to the axis of rotation. One should take caution not to evaluate the square of the difference of the two functions, but to evaluate the difference of the squares of the two functions.

$$R_O(x)^2 - R_I(x)^2 \not\equiv \left(R_O(x) - R_I(x)\right)^2$$

(This formula only works for revolutions about the x-axis.)

To rotate about any horizontal axis, simply subtract from that axis each formula. If h is the value of a horizontal axis, then the volume equals,

$$\pi\int_a^b \left(\left(h - R_O(x)\right)^2 - \left(h - R_I(x)\right)^2\right)dx.$$

For example, to rotate the region between $y = -2x + x^2$ and $y = x$ along the axis $y = 4$, one would integrate as follows:

$$\pi\int_0^3 \left(\left(4 - \left(-2x + x^2\right)\right)^2 - (4 - x)^2\right)dx.$$

The bounds of integration are the zeros of the first equation minus the second. Note that when integrating along an axis other than the x, the further axis may not be that obvious. In the previous example, even though $y = x$ is further up than $y = -2x + x^2$, it is the inner axis since it is closer to, $y = 4$

The same idea can be applied to both the y-axis and any other vertical axis. One simply must solve each equation for x before one inserts them into the integration formula.

Shell Integration

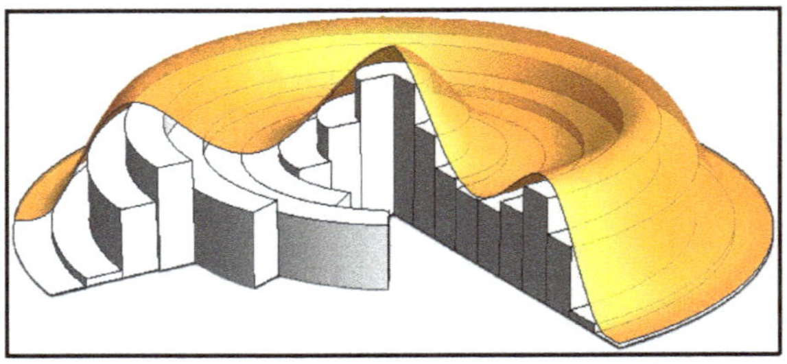

A volume is approximated by a collection of hollow cylinders. As the cylinders get smaller the approximation gets better. The limit of this approximation is the shell integral.

Shell integration (the shell method in integral calculus) is a means of calculating the volume of a solid of revolution, when integrating along an axis *perpendicular to* the axis of revolution. This is in contrast to disk integration which integrates along the axis *parallel* to the axis of revolution.

Definition

The shell method goes as follows: Consider a volume in three dimensions obtained by rotating a cross-section in the *xy*-plane around the *y*-axis. Suppose the cross-section is defined by the graph of the positive function *f*(*x*) on the interval [*a*, *b*]. Then the formula for the volume will be:

$$2\pi\int_a^b xf(x)dx$$

If the function is of the *y* coordinate and the axis of rotation is the *x*-axis then the formula becomes:

$$2\pi\int_a^b yf(y)dy$$

If the function is rotating around the line *x* = *h* or *y* = *k*, the formulas become:

$$\begin{cases} 2\pi\int_a^b (x-h)f(x)dx, & \text{if } h \le a < b \\ 2\pi\int_a^b (h-x)f(x)dx, & \text{if } a < b \le h \end{cases}$$

and

$$\begin{cases} 2\pi\int_a^b (y-k)f(y)dx, & \text{if } k \le a < b \\ 2\pi\int_a^b (k-y)f(y)dx, & \text{if } a < b \le k \end{cases}$$

The formula is derived by computing the double integral in polar coordinates.

Example

Consider the volume, depicted below, whose cross section on the interval [1, 2] is defined by:

$$y=(x-1)^2(x-2)^2$$

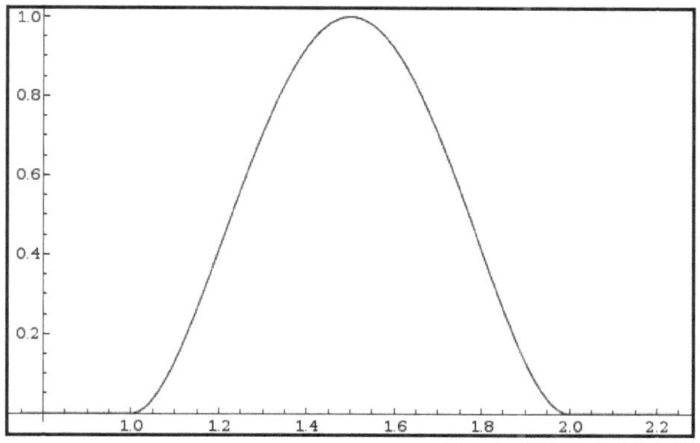

Cross-section.

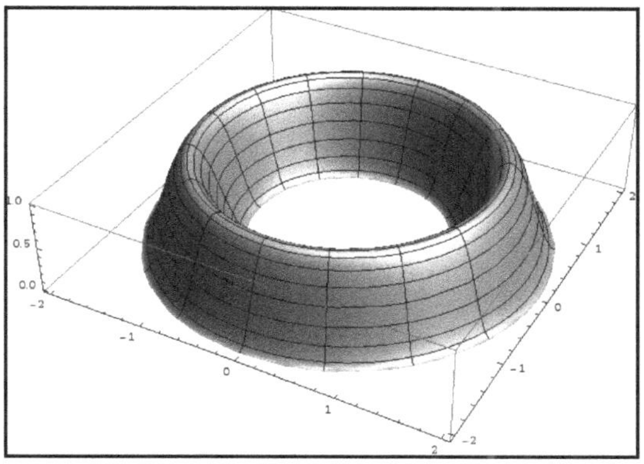

3D volume.

In the case of disk integration we would need to solve for x given y. Because the volume is hollow in the middle we will find two functions, one that defines the inner solid and one that defines the outer solid. After integrating these two functions with the disk method we subtract them to yield the desired volume.

With the shell method all we need is the following formula:

$$2\pi\int_{1}^{2} x(x-1)^2(x-2)^2\,dx$$

By expanding the polynomial the integral becomes very simple. In the end we find the volume is $\pi/10$ cubic units.

Integration by Substitution

In calculus, integration by substitution, also known as u-substitution, is a method for finding integrals. Using the fundamental theorem of calculus often requires finding an antiderivative. For this and other reasons, integration by substitution is an important tool in mathematics. It is the counterpart to the chain rule of differentiation.

Substitution for Single Variable

Definition

Let $I \subseteq R$ be an interval and $\phi : [a,b] \to I$ be a differentiable function with integrable derivative. Suppose that $f : I \to R$ is a continuous function. Then,

$$\int_{\varphi(a)}^{\varphi(b)} f(x)dx = \int_a^b f(\varphi(t))\varphi'(t)dt.$$

In other notation: the substitution $x = \phi(t)$ yields $dx/dt = \phi'(t)$ and thus, formally, $dx = \phi'(t)\ dt$, which is the required substitution for dx. (One could view the method of integration by substitution as a justification of Leibniz's notation for integrals and derivatives.)

The formula is used to transform one integral into another integral that is easier to compute. Thus, the formula can be used from left to right or from right to left in order to simplify a given integral. When used in the latter manner, it is sometimes known as **u-substitution** or **w-substitution**.

Proof

Integration by substitution can be derived from the fundamental theorem of calculus as follows. Let f and ϕ be two functions satisfying the above hypothesis that f is continuous on I and ϕ' is integrable on the closed interval $[a,b]$. Then the function $f(\phi(t))\phi'(t)$ is also integrable on $[a,b]$. Hence the integrals:

$$\int_{\varphi(a)}^{\varphi(b)} f(x)dx$$

and

$$\int_a^b f(\varphi(t))\varphi'(t)dt$$

in fact exist, and it remains to show that they are equal.

Since f is continuous, it has an antiderivative F. The composite function $F \circ \phi$ is then defined. Since F and ϕ are differentiable, the chain rule gives:

$$(F^\circ\varphi)'(t) = F'(\varphi(t))\varphi'(t) = f(\varphi(t))\varphi'(t).$$

Applying the fundamental theorem of calculus twice gives:

$$\begin{aligned}
&\int_a^b f(\varphi(t))\varphi'(t)dt = \int_a^b (F^\circ\varphi)'(t)dt \\
&= (F^\circ\varphi)(b) - (F^\circ\varphi)(a) \\
&= F(\varphi(b)) - F(\varphi(a)) \\
&= \int_{\varphi(a)}^{\varphi(b)} f(x)dx,
\end{aligned}$$

which is the substitution rule.

Examples

Example 1: from Right to Left

Consider the integral:

$$\int_0^2 x\cos(x^2+1)dx$$

If we apply the formula from right to left and make the substitution $u = \phi(x) = x^2 + 1$, we obtain $du = 2x\,dx$ and hence; $x\,dx = ½du$

$$\int_{x=0}^{x=2} x\cos(x^2+1)dx = \frac{1}{2}\int_{u=1}^{u=5}\cos(u)du$$

$$= \frac{1}{2}(\sin(5)-\sin(1)).$$

It is important to note that since the lower limit $x = 0$ was replaced with $u = 0^2 + 1 = 1$, and the upper limit $x = 2$ replaced with $u = 2^2 + 1 = 5$, a transformation back into terms of x was unnecessary.

Example 2: from Left to Right

For the integral:

$$\int_0^1 \sqrt{1-x^2}\;dx$$

the formula needs to be used from left to right: the substitution $x = \sin(u)$, $dx = \cos(u)\,du$ is useful, because $\sqrt{(1-\sin^2(u))} = \cos(u)$:

$$\int_0^1 \sqrt{1-x^2}\;dx = \int_0^{\frac{\pi}{2}} \sqrt{1-\sin^2(u)}\cos(u)\;du = \int_0^{\frac{\pi}{2}} \cos^2(u)\;du = \frac{\pi}{4}$$

The resulting integral can be computed using integration by parts or a double angle formula followed by one more substitution. One can also note that the function being integrated is the upper right quarter of a circle with a radius of one, and hence integrating the upper right quarter from zero to one is the geometric equivalent to the area of one quarter of the unit circle, or $\pi/4$.

Example 3: Antiderivatives

Substitution can be used to determine antiderivatives. One chooses a relation between x and u, determines the corresponding relation between dx and du by differentiating, and performs the substitutions. An antiderivative for the substituted function can hopefully be determined; the original substitution between u and x is then undone.

Similar to our first example above, we can determine the following antiderivative with this method:

$$\int x\cos(x^2+1)dx = \frac{1}{2}\int 2x\cos(x^2+1)dx = \frac{1}{2}\int \cos u\,du$$

$$= \frac{1}{2}\sin u + C = \frac{1}{2}\sin(x^2+1) + C$$

where C is an arbitrary constant of integration.

Note that there were no integral boundaries to transform, but in the last step we had to revert the original substitution $u = x^2 + 1$.

Substitution for Multiple Variables

One may also use substitution when integrating functions of several variables. Here the substitution function $(v_1,\ldots,v_n) = \varphi(u_1, \ldots, u_n)$ needs to be injective and continuously differentiable, and the differentials transform as:

$$dv_1 \cdots dv_n = |\det(\mathrm{D}\varphi)(u_1,\ldots,u_n)|\, du_1 \cdots du_n$$

where $\det(\mathrm{D}\varphi)(u_1, \ldots, u_n)$ denotes the determinant of the Jacobian matrix containing the partial derivatives of φ. This formula expresses the fact that the absolute value of the determinant of a matrix equals the volume of the parallelotope spanned by its columns or rows.

More precisely, the *change of variables* formula is stated in the next theorem:

Theorem. Let U be an open set in R^n and $\varphi : U \to \mathrm{R}^n$ an injective differentiable function with continuous partial derivatives, the Jacobian of which is nonzero for every x in U. Then for any real-valued, compactly supported, continuous function f, with support contained in $\varphi(U)$,

$$\int_{\varphi(U)} f(\mathbf{v})\,d\mathbf{v} = \int_U f(\varphi(\mathbf{u}))\left|\det(\mathrm{D}\varphi)(\mathbf{u})\right| d\mathbf{u}.$$

The conditions on the theorem can be weakened in various ways. First, the requirement that φ be continuously differentiable can be replaced by the weaker assumption that φ be merely differentiable and have a continuous inverse (Rudin 1987, Theorem 7.26). This is guaranteed to hold if φ is continuously differentiable by the inverse function theorem. Alternatively, the requirement that $\mathrm{Det}(\mathrm{D}\varphi)\neq 0$ can be eliminated by applying Sard's theorem (Spivak 1965).

For Lebesgue measurable functions, the theorem can be stated in the following form (Fremlin 2010, Theorem 263D):

Theorem. Let U be a measurable subset of R^n and $\varphi: U \to \mathrm{R}^n$ an injective function, and suppose for every x in U there exists $\varphi'(x)$ in $\mathrm{R}^{n,n}$ such that $\varphi(y) = \varphi(x) + \varphi'(x)(y - x) + o(||y - x||)$ as $y \to x$. Then $\varphi(U)$ is measurable, and for any real-valued function f defined on $\varphi(U)$,

$$\int_{\varphi(U)} f(v)\,dv = \int_U f(\varphi(u))\left|\det \varphi'(u)\right| du$$

in the sense that if either integral exists (or is properly infinite), then so does the other one, and they have the same value.

Another very general version in measure theory is the following (Hewitt & Stromberg 1965, Theorem 20.3):

Theorem. Let X be a locally compact Hausdorff space equipped with a finite Radon measure μ, and let Y be a σ-compact Hausdorff space with a σ-finite Radon measure ρ. Let $\varphi : X \to Y$ be a continuous and absolutely continuous function (where the latter means that $\rho(\varphi(E)) = 0$ whenever $\mu(E) = 0$). Then there exists a real-valued Borel measurable function w on X such that for every Lebesgue integrable function $f: Y \to \mathrm{R}$, the function $(f \circ \varphi)w$ is Lebesgue integrable on X, and,

$$\int_Y f(y)d\rho(y) = \int_X f\circ\varphi(x)w(x)d\mu(x).$$

Furthermore, it is possible to write,

$$w(x) = g\circ\varphi(x)$$

for some Borel measurable function g on Y.

In geometric measure theory, integration by substitution is used with Lipschitz functions. A bi-Lipschitz function is a Lipschitz function $\varphi : U \to \mathrm{R}^n$ which is injective, and such that its inverse function $\varphi^{-1} : \varphi(U) \to U$ is also Lipschitz. By Rademacher's theorem a bi-Lipschitz mapping is differentiable almost everywhere. In particular, the Jacobian determinant of a bi-Lipschitz mapping det Dφ is well-defined almost everywhere. The following result then holds:

Theorem. Let U be an open subset of R^n and $\varphi : U \to \mathrm{R}^n$ be a bi-Lipschitz mapping. Let $f : \varphi(U) \to$ **R** be measurable. Then,

$$\int_U (f\circ\varphi)\,|\det D\varphi| = \int_{\varphi(U)} f$$

in the sense that if either integral exists (or is properly infinite), then so does the other one, and they have the same value.

The above theorem was first proposed by Euler when he developed the notion of double integrals in 1769. Although generalized to triple integrals by Lagrange in 1773, and used by Legendre, Laplace, Gauss, and first generalized to n variables by Mikhail Ostrogradski in 1836, it resisted a fully rigorous formal proof for a surprisingly long time, and was first satisfactorily resolved 125 years later, by Élie Cartan in a series of papers beginning in the mid-1890s (Katz 1982; Ferzola 1994).

Application in Probability

Substitution can be used to answer the following important question in probability: given a random variable X with probability density p_x and another random variable Y related to X by the equation $y = \varphi(x)$, what is the probability density for Y?

It is easiest to answer this question by first answering a slightly different question: what is the probability that Y takes a value in some particular subset S? Denote this probability $P(Y \in S)$. Of course, if Y has probability density p_y then the answer is:

$$P(Y \in S) = \int_S p_y(y)dy,$$

but this isn't really useful because we don't know p_y; it's what we're trying to find in the first place. We can make progress by considering the problem in the variable X. Y takes a value in S whenever X takes a value in $\phi^{-1}(S)$, so,

$$P(Y \in S) = \int_{\phi^{-1}(S)} p_x(x)dx.$$

Changing from variable x to y gives,

$$P(Y \in S) = \int_{\phi^{-1}(S)} p_x(x)dx = \int_S p_x(\phi^{-1}(y))\left|\frac{d\phi^{-1}}{dy}\right|dy.$$

Combining this with our first equation gives,

$$\int_S p_y(y)dy = \int_S p_x(\phi^{-1}(y))\left|\frac{d\phi^{-1}}{dy}\right|dy$$

so,

$$p_y(y) = p_x(\phi^{-1}(y))\left|\frac{d\phi^{-1}}{dy}\right|.$$

In the case where X and Y depend on several uncorrelated variables, i.e. $p_x = p_x(x_1 \ldots x_n)$, and $y = \phi(x)$, p_y can be found by substitution in several variables discussed above. The result is:

$$p_y(y) = p_x(\phi^{-1}(y))\left|\det\left[D\phi^{-1}(y)\right]\right|.$$

Trigonometric Substitution

In mathematics, Trigonometric substitution is the substitution of trigonometric functions for other expressions. One may use the trigonometric identities to simplify certain integrals containing radical expressions:

Substitution 1. If the integrand contains $a^2 - x^2$, let,

$$x = a \sin\theta$$

and use the identity,

$$1 - \sin^2\theta = \cos^2\theta.$$

Substitution 2. If the integrand contains $a^2 + x^2$, let,

$$x = a \tan\theta$$

and use the identity,

$$1 + \tan^2\theta = \sec^2\theta.$$

Substitution 3. If the integrand contains $x^2 - a^2$, let,

$$x = a \sec\theta$$

and use the identity,

$$\sec^2\theta - 1 = \tan^2\theta.$$

Examples

Integrals Containing $a^2 - x^2$

In the integral,

$$\int \frac{dx}{\sqrt{a^2 - x^2}}$$

we may use,

$$x = a\sin(\theta), \quad \mathrm{d}x = a\cos(\theta)\mathrm{d}\theta, \quad \theta = \arcsin\left(\frac{x}{a}\right)$$

$$\begin{aligned}
\int \frac{\mathrm{d}x}{\sqrt{a^2 - x^2}} &= \int \frac{a\cos(\theta)\mathrm{d}\theta}{\sqrt{a^2 - a^2\sin^2(\theta)}} \\
&= \int \frac{a\cos(\theta)\mathrm{d}\theta}{\sqrt{a^2(1-\sin^2(\theta))}} \\
&= \int \frac{a\cos(\theta)\mathrm{d}\theta}{\sqrt{a^2\cos^2(\theta)}} \\
&= \int \mathrm{d}\theta \\
&= \theta + C \\
&= \arcsin\left(\tfrac{x}{a}\right) + C
\end{aligned}$$

Note that the above step requires that $a > 0$ and $\cos(\theta) > 0$; we can choose the a to be the positive square root of a^2; and we impose the restriction on θ to be $-\pi/2 < \theta < \pi/2$ by using the arcsin function.

For a definite integral, one must figure out how the bounds of integration change. For example, as x goes from 0 to $a/2$, then $\sin(\theta)$ goes from 0 to $1/2$, so θ goes from 0 to $\pi/6$. Then we have,

$$\int_0^{\frac{a}{2}} \frac{\mathrm{d}x}{\sqrt{a^2 - x^2}} = \int_0^{\frac{\pi}{6}} \mathrm{d}\theta = \tfrac{\pi}{6}.$$

Some care is needed when picking the bounds. The integration above requires that $-\pi/2 < \theta < \pi/2$, so θ going from 0 to $\pi/6$ is the only choice. If we had missed this restriction, we might have picked θ to go from π to $5\pi/6$, which would give us the negative of the result.

Integrals Containing $a^2 + x^2$

In the integral,

$$\int \frac{\mathrm{d}x}{a^2 + x^2}$$

we may write,

$$x = a\tan(\theta), \quad \mathrm{d}x = a\sec^2(\theta)\mathrm{d}\theta, \quad \theta = \arctan\left(\tfrac{x}{a}\right)$$

so that the integral becomes,

$$\int \frac{dx}{a^2 + x^2} = \int \frac{a\sec^2(\theta)d\theta}{a^2 + a^2\tan^2(\theta)}$$
$$= \int \frac{a\sec^2(\theta)d\theta}{a^2(1+\tan^2(\theta))}$$
$$= \int \frac{a\sec^2(\theta)d\theta}{a^2\sec^2(\theta)}$$
$$= \int \frac{d\theta}{a}$$
$$= \tfrac{\theta}{a} + C$$
$$= \tfrac{1}{a}\arctan\left(\tfrac{x}{a}\right) + C$$

(provided $a \neq 0$).

Integrals Containing $x^2 - a^2$

Integrals like,

$$\int \frac{dx}{x^2 - a^2}$$

should be done by partial fractions rather than trigonometric substitutions. However, the integral,

$$\int \sqrt{x^2 - a^2}\,dx$$

can be done by substitution:

$$x = a\sec(\theta), \quad dx = a\sec(\theta)\tan(\theta)d\theta, \quad \theta = arcsec\left(\frac{x}{a}\right)$$

$$\int \sqrt{x^2 - a^2}\,dx = \int \sqrt{a^2\sec^2(\theta) - a^2} \cdot a\sec(\theta)\tan(\theta)d\theta$$
$$= \int \sqrt{a^2(\sec^2(\theta) - 1)} \cdot a\sec(\theta)\tan(\theta)d\theta$$
$$= \int \sqrt{a^2\tan^2(\theta)} \cdot a\sec(\theta)\tan(\theta)d\theta$$
$$= \int a^2\sec(\theta)\tan^2(\theta)d\theta$$
$$= a^2\int \sec(\theta)(\sec^2(\theta) - 1)d\theta$$
$$= a^2\int (\sec^3(\theta) - \sec(\theta))d\theta.$$

We can then solve this using the formula for the integral of secant cubed.

Substitutions that Eliminate Trigonometric Functions

Substitution can be used to remove trigonometric functions.

For instance,

$$\int f(\sin(x),\cos(x))\mathrm{d}x = \int \frac{1}{\pm\sqrt{1-u^2}} f\left(u, \pm\sqrt{1-u^2}\right)\mathrm{d}u = \left(u = \sin(x)\right)$$

$$\int f(\sin(x),\cos(x))\mathrm{d}x = \int \frac{1}{\mp\sqrt{1-u^2}} f\left(\pm\sqrt{1-u^2}, u\right)\mathrm{d}u = \left(u = \cos(x)\right)$$

$$\int f(\sin(x),\cos(x))\mathrm{d}x = \int \frac{2}{1+u^2} f\left(\frac{2u}{1+u^2}, \frac{1-u^2}{1+u^2}\right)\mathrm{d}u = \left(u = \tan\left(\tfrac{x}{2}\right)\right)$$

$$\int \frac{\cos x}{(1+\cos x)^3}\mathrm{d}x = \int \frac{2}{1+u^2} \frac{\frac{1-u^2}{1+u^2}}{\left(1+\frac{1-u^2}{1+u^2}\right)^3}\mathrm{d}u = \int \frac{1-u^2}{1+u^2}\mathrm{d}u$$

Hyperbolic Substitution

Substitutions of hyperbolic functions can also be used to simplify integrals.

In the integral $\int \frac{1}{\sqrt{a^2+x^2}}\mathrm{d}x$, make the substitution $x = a\sinh u$, $\mathrm{d}x = a\cosh u\,\mathrm{d}u$.

Then, using the identities $\cosh^2(x) - \sinh^2(x) = 1$ and $\sinh^{-1} x = \ln(x + \sqrt{x^2+1})$,

$$\begin{aligned}
\int \frac{1}{\sqrt{a^2+x^2}}\mathrm{d}x &= \int \frac{a\cosh u}{\sqrt{a^2+a^2\sinh^2 u}}\mathrm{d}u \\
&= \int \frac{a\cosh u}{a\sqrt{1+\sinh^2 u}}\mathrm{d}u \\
&= \int \frac{a\cosh u}{a\cosh u}\mathrm{d}u \\
&= u + C \\
&= \sinh^{-1}\frac{x}{a} + C \\
&= \ln\left(\sqrt{\frac{x^2}{a^2}+1} + \frac{x}{a}\right) + C \\
&= \ln\left(\frac{\sqrt{x^2+a^2}+x}{a}\right) + C
\end{aligned}$$

Order of Integration

In calculus, interchange of the order of integration is a methodology that transforms iterated integrals (or multiple integrals through the use of Fubini's theorem) of functions into other, hopefully simpler, integrals by changing the order in which the integrations are performed. In some cases, the order of integration can be validly interchanged; in others it cannot.

Problem Statement

The problem for examination is evaluation of an integral of the form:

$$\iint_D f(x,y)\,dxdy,$$

where D is some two-dimensional area in the xy–plane. For some functions f straightforward integration is feasible, but where that is not true, the integral can sometimes be reduced to simpler form by changing the order of integration. The difficulty with this interchange is determining the change in description of the domain D.

The method also is applicable to other multiple integrals.

Sometimes, even though a full evaluation is difficult, or perhaps requires a numerical integration, a double integral can be reduced to a single integration, as illustrated next. Reduction to a single integration makes a numerical evaluation much easier and more efficient.

Relation to Integration by Parts

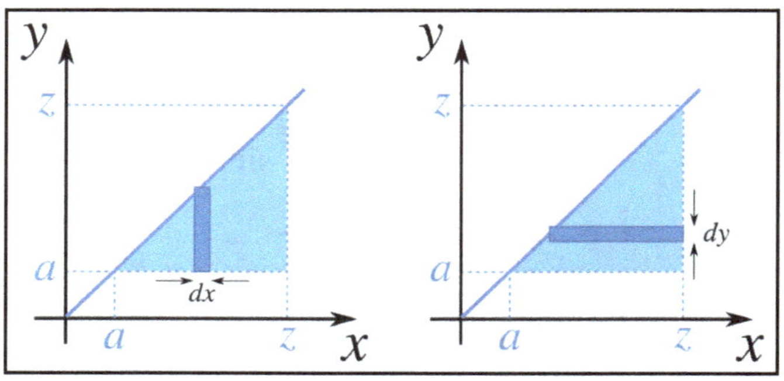

Figure 1: Integration over the triangular area can be done using vertical or horizontal strips as the first step. This is an overhead view, looking down the z-axis onto the x-y plane. The sloped line is the curve $y = x$.

Consider the iterated integral:

$$\int_a^z \int_a^x h(y)dydx\,,,$$

which we will write using the prefix notation commonly seen in physics:

$$\int_a^z dx \int_a^x h(y)dy.$$

In this expression, the second integral is calculated first with respect to y and x is held constant—a strip of width dx is integrated first over the y-direction (a strip of width dx in the x direction is integrated with respect to the y variable across the y direction), adding up an infinite amount of rectangles of width dy along the y-axis. This forms a three dimensional slice dx wide along the x-axis, from y=a to y=x along the y axis, and in the z direction z=f(x,y). Notice that if the thickness dx is infinitesimal, x varies only infinitesimally on the slice. We can assume that x is constant. This integration is as shown in the left panel of Figure 1, but is inconvenient especially when the function $h(y)$ is not easily integrated. The integral can be reduced to a single integration by reversing the order of integration as shown in the right panel of the figure. To accomplish this interchange of variables, the strip of width dy is first integrated from the line $x = y$ to the limit $x = z$, and then the result is integrated from $y = a$ to $y = z$, resulting in:

$$\int_a^z dx \int_a^x h(y)\, dy = \int_a^z h(y)\, dy \int_y^z dx = \int_a^z (z-y)h(y)dy\,.$$

This result can be seen to be an example of the formula for integration by parts, as stated below:

$$\int_a^z f(x)g'(x)dx = [f(x)g(x)]_a^z - \int_a^z f'(x)g(x)dx$$

Substitute:

$$f(x) = \int_a^x h(y)dy \text{ and } g'(x) = 1\,.$$

Which gives the result.

Principal-value Integrals

For application to principal-value integrals, see Whittaker and Watson, Gakhov, Lu, or Zwillinger. An example where the order of integration cannot be exchanged is given by Kanwal:

$$\frac{1}{(2\pi i)^2}\int_L^* \frac{d\tau_1}{\tau_1 - t}\int_L^* g(\tau)\frac{d\tau}{\tau-\tau_1} = \frac{1}{4}g(t)\,,$$

while:

$$\frac{1}{(2\pi i)^2}\int_L^* g(\tau)\,d\tau\left(\int_L^* \frac{d\tau_1}{(\tau_1 - t)(\tau-\tau_1)}\right) = 0\,.$$

The second form is evaluated using a partial fraction expansion and an evaluation using the Sokhotski–Plemelj formula:

$$\int_L^* \frac{d\tau_1}{\tau_1 - t} = \int_L^* \frac{d\tau_1}{\tau_1 - t} = \pi\, i\,.$$

The notation $\int_L^*$ indicates a Cauchy principal value.

Basic Theorems

A good discussion of the basis for reversing the order of integration is found in the book *Fourier Analysis* by T.W. Körner. He introduces his discussion with an example where interchange of integration leads to two different answers because the conditions of Theorem II below are not satisfied. Here is the example:

$$\int_1^\infty \frac{x^2-y^2}{\left(x^2+y^2\right)^2}\,dy = \left[\frac{y}{x^2+y^2}\right]_1^\infty = -\frac{1}{1+x^2}\,[x \geq 1]\,.$$

$$\int_1^\infty \left(\int_1^\infty \frac{x^2 - y^2}{\left(x^2 + y^2\right)^2} dy \right) dx = -\frac{\pi}{4}.$$

$$\int_1^\infty \left(\int_1^\infty \frac{x^2 - y^2}{\left(x^2 + y^2\right)^2} dx \right) dy = \frac{\pi}{4}.$$

Two basic theorems governing admissibility of the interchange are quoted below from Chaudhry and Zubair:

Theorem I — Let $f(x, y)$ be a continuous function of constant sign defined for $a \le x < \infty$, $c \le y < \infty$, and let the integrals

$$J(y) := \int_a^\infty dx\, f(x, y) \qquad \text{and} \qquad J^*(x) = \int_c^\infty dy\, f(x, y)$$

regarded as functions of the corresponding parameter be, respectively, continuous for $c \le y < \infty$, $a \le x < \infty$. Then if at least one of the iterated integrals:

$$\int_c^\infty dy \left(\int_a^\infty dx\, f(x, y) \right) \qquad \text{and} \qquad \int_a^\infty dx \left(\int_c^\infty dy\, f(x, y) \right)$$

converges, the other integral also converges and their values coincide.

Theorem II — Let $f(x, y)$ be continuous for $a \le x < \infty$, $c \le y < \infty$, and let the integrals:

$$J(y) := \int_a^\infty dx\, f(x, y) \qquad \text{and} \qquad J^*(x) = \int_c^\infty dy\, f(x, y)$$

be respectively, uniformly convergent on every finite interval $c \le y < C$ and on every finite interval $a \le x < A$. Then if at least one of the iterated integrals:

$$\int_c^\infty dy \left(\int_a^\infty dx\, |\, f(x, y)\,| \right) \qquad \text{and} \qquad \int_a^\infty dx \left(\int_c^\infty dy\, |\, f(x, y)\,| \right)$$

converges, the iterated integrals:

$$\int_c^\infty dy \left(\int_a^\infty dx\, f(x, y) \right) \qquad \text{and} \qquad \int_a^\infty dx \left(\int_c^\infty dy\, f(x, y) \right)$$

also converge and their values are equal.

The most important theorem for the applications is quoted from Protter and Morrey:

Theorem — Suppose F is a region given by $F = \{(x, y) : a \le x \le b, p(x) \le y \le q(x)\}$ where p and q are continuous and $p(x) \le q(x)$ for $a \le x \le b$. Suppose that $f(x, y)$ is continuous on F. Then,

$$\iint_F f(x, y)dA = \int_a^b \int_{p(x)}^{q(x)} f(x, y)dy\, dx.$$

The corresponding result holds if the closed region F has the representation $F = \{(x, y) : c \le y \le d, r(y) \le x \le s(y)\}$ where $r(y) \le s(y)$ for $c \le y \le d$. In such a case,

$$\iint_F f(x, y)dA = \int_c^d \int_{r(y)}^{s(y)} f(x, y)dx\, dy.$$

In other words, both iterated integrals, when computable, are equal to the double integral and therefore equal to each other.

Integration by Reduction Formulae

Integration by reduction formula in integral calculus is a technique of integration, in the form of a recurrence relation. It is used when an expression containing an integer parameter, usually in the form of powers of elementary functions, or products of transcendental functions and polynomials of arbitrary degree, can't be integrated directly. But using other methods of integration a reduction formula can be set up to obtain the integral of the same or similar expression with a lower integer parameter, progressively simplifying the integral until it can be evaluated. This method of integration is one of the earliest used.

How to Find the Reduction Formula

The reduction formula can be derived using any of the common methods of integration, like integration by substitution, integration by parts, integration by trigonometric substitution, integration by partial fractions, etc. The main idea is to express an integral involving an integer parameter (e.g. power) of a function, represented by I_n, in terms of an integral that involves a lower value of the parameter (lower power) of that function, for example I_{n-1} or I_{n-2}. This makes the reduction formula a type of recurrence relation. In other words, the reduction formula expresses the integral,

$$I_n = \int f(x,n)\mathrm{d}x,$$

in terms of,

$$I_k = \int f(x,k)\mathrm{d}x,$$

where,

$$k < n.$$

How to Compute the Integral

To compute the integral, we set n to its value and use the reduction formula to calculate the $(n - 1)$ or $(n - 2)$ integral. The higher index integral can be used to calculate lower index ones; the process is continued repeatedly until we reach a point where the function to be integrated can be computed, usually when its index is 0 or 1. Then we back-substitute the previous results until we have computed I_n.

Examples

Below are examples of the procedure.

Cosine Integral

Typically, integrals like, $\int \cos^n x\mathrm{d}x$,

can be evaluated by a reduction formula.

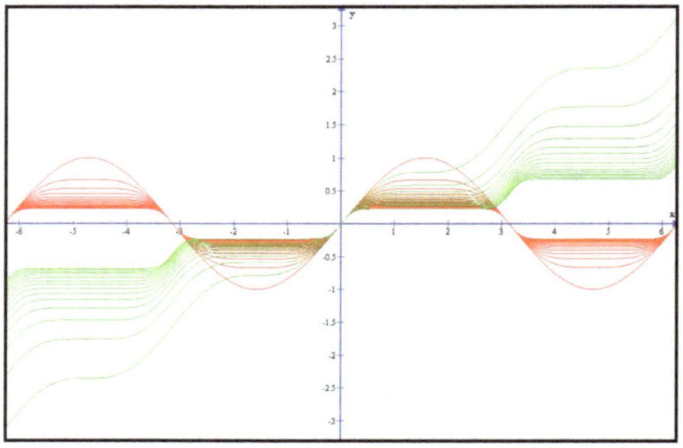

$\int \cos^n(x)\,dx$, for $n = 1, 2 \ldots 30$

Start by setting:

$$I_n = \int \cos^n x \mathrm{d}x.$$

Now re-write as:

$$I_n = \int \cos^{n-1} x \cos x \mathrm{d}x,$$

Integrating by this substitution:

$$\cos x \mathrm{d}x = \mathrm{d}(\sin x),$$

$$I_n = \int \cos^{n-1} x \mathrm{d}(\sin x).$$

Now integrating by parts:

$$\begin{aligned}\int \cos^n x \mathrm{d}x &= \cos^{n-1} x \sin x - \int \sin x \mathrm{d}(\cos^{n-1} x) \\ &= \cos^{n-1} x \sin x + (n-1)\int \sin x \cos^{n-2} x \sin x \mathrm{d}x \\ &= \cos^{n-1} x \sin x + (n-1)\int \cos^{n-2} x \sin^2 x \mathrm{d}x \\ &= \cos^{n-1} x \sin x + (n-1)\int \cos^{n-2} x(1-\cos^2 x)\mathrm{d}x \\ &= \cos^{n-1} x \sin x + (n-1)\int \cos^{n-2} x \mathrm{d}x - (n-1)\int \cos^n x \mathrm{d}x \\ &= \cos^{n-1} x \sin x + (n-1)I_{n-2} - (n-1)I_n,\end{aligned}$$

solving for I_n:

$$I_n + (n-1)I_n = \cos^{n-1} x \sin x + (n-1)I_{n-2},$$

$$nI_n = \cos^{n-1}(x) \sin x + (n-1)I_{n-2},$$

$$I_n = \frac{1}{n}\cos^{n-1}x\sin x + \frac{n-1}{n}I_{n-2},$$

so the reduction formula is:

$$\int \cos^n x\mathrm{d}x = \frac{1}{n}\cos^{n-1}x\sin x + \frac{n-1}{n}\int \cos^{n-2}x\mathrm{d}x.$$

To supplement the example, the above can be used to evaluate the integral for (say) $n = 5$;

$$I_5 = \int \cos^5 x\mathrm{d}x.$$

Calculating lower indices:

$$n = 5,\quad I_5 = \tfrac{1}{5}\cos^4 x\sin x + \tfrac{4}{5}I_3,$$

$$n = 3,\quad I_3 = \tfrac{1}{3}\cos^2 x\sin x + \tfrac{2}{3}I_1,$$

back-substituting:

$$\because I_1 = \int \cos x\mathrm{d}x = \sin x + C_1,$$

$$\therefore I_3 = \tfrac{1}{3}\cos^2 x\sin x + \tfrac{2}{3}\sin x + C_2,\quad C_2 = \tfrac{2}{3}C_1,$$

$$I_5 = \frac{1}{5}\cos^4 x\sin x + \frac{4}{5}\left[\frac{1}{3}\cos^2 x\sin x + \frac{2}{3}\sin x\right] + C,$$

where C is a constant.

Exponential Integral

Another typical example is:

$$\int x^n e^{ax}\,\mathrm{d}x..$$

Start by setting:

$$I_n = \int x^n e^{ax}\,\mathrm{d}x.$$

Integrating by substitution:

$$x^n\,\mathrm{d}x = \frac{\mathrm{d}(x^{n+1})}{n+1},$$

$$I_n = \frac{1}{n+1}\int e^{ax}\,\mathrm{d}(x^{n+1}),$$

Now integrating by parts:

$$\int e^{ax}\,\mathrm{d}(x^{n+1}) = x^{n+1}e^{ax} - \int x^{n+1}\,\mathrm{d}(e^{ax})$$

$$= x^{n+1}e^{ax} - a\int x^{n+1}e^{ax}\,\mathrm{d}x,$$

$$(n+1)I_n = x^{n+1}e^{ax} - aI_{n+1},$$

shifting indices back by 1 (so $n + 1 \to n, n \to n - 1$):

$$nI_{n-1} = x^n e^{ax} - aI_n,$$

solving for *In*:

$$I_n = \frac{1}{a}\left(x^n e^{ax} - nI_{n-1}\right),$$

so the reduction formula is:

$$\int x^n e^{ax}\,\mathrm{d}x = \frac{1}{a}\left(x^n e^{ax} - n\int x^{n-1} e^{ax}\,\mathrm{d}x\right).$$

Tables of Integral Reduction Formulas

Rational Functions

The following integrals contain:

- Factors of the linear radical $\sqrt{ax+b}$.
- Linear factors $px+q$ and the linear radical $\sqrt{ax+b}$.
- Quadratic factors x^2+a^2.
- Quadratic factors x^2-a^2 , for $x>a$.
- Quadratic factors a^2-x^2 , for $x<a$.
- (Irreducible) quadratic factors ax^2+bx+c.
- Radicals of irreducible quadratic factors $\sqrt{ax^2+bx+c}$.

Integral	Reduction formula
$I_n = \int \frac{x^n}{\sqrt{ax+b}}\mathrm{d}x$	$I_n = \frac{2x^n\sqrt{ax+b}}{a(2n+1)} - \frac{2nb}{a(2n+1)}I_{n-1}$
$I_n = \int \frac{\mathrm{d}x}{x^n\sqrt{ax+b}}$	$I_n = -\frac{\sqrt{ax+b}}{(n-1)bx^{n-1}} - \frac{a(2n-3)}{2b(n-1)}I_{n-1}$
$I_n = \int x^n\sqrt{ax+b}\,\mathrm{d}x$	$I_n = \frac{2x^n\sqrt{(ax+b)^3}}{a(2n+3)} - \frac{2nb}{a(2n+3)}I_{n-1}$
$I_{n,m} = \int \frac{\mathrm{d}x}{(ax+b)^n(px+q)^m}$	$I_{n,m} = -\frac{1}{(n-1)(bp-aq)}\left[\frac{1}{(ax+b)^{m-1}(px+q)^{n-1}} + a(n+m-2)I_{m,n-1}\right]$

$I_{n,m}=\int\frac{(ax+b)^m}{(px+q)^n}\mathrm{d}x$	$I_{n,m}=\begin{cases}-\frac{1}{(n-1)(bp-aq)}\left[\frac{(ax+b)^{m+1}}{(px+q)^{n-1}}+a(n+m-2)I_{m-1,n-1}\right]\\-\frac{1}{(n-m-1)p}\left[\frac{(ax+b)^m}{(px+q)^{n-1}}+m(bp-aq)I_{m-1,n}\right]\\-\frac{1}{(n-1)p}\left[\frac{(ax+b)^m}{(px+q)^{n-1}}-amI_{m-1,n-1}\right]\end{cases}$

Integral	Reduction formula
$I_n=\int\frac{(px+q)^n}{\sqrt{ax+b}}\mathrm{d}x$	$\int(px+q)^n\sqrt{ax+b}\,\mathrm{d}x=\frac{2(px+q)^{n+1}\sqrt{ax+b}}{p(2n+3)}+\frac{bp-aq}{p(2n+3)}I_n$ $I_n=\frac{2(px+q)^n\sqrt{ax+b}}{a(2n+1)}+\frac{2n(aq-bp)}{a(2n+1)}I_{n-1}$
$I_n=\int\frac{\mathrm{d}x}{(px+q)^n\sqrt{ax+b}}$	$\int\frac{\sqrt{ax+b}}{(px+q)^n}\mathrm{d}x=-\frac{\sqrt{ax+b}}{p(n-1)(px+q)^{n-1}}-\frac{a}{2p(n-1)}I_n$ $I_n=-\frac{\sqrt{ax+b}}{(n-1)(aq-bp)(px+q)^{n-1}}+\frac{a(2n-3)}{2(n-1)(aq-bp)}I_{n-1}$

Integral	Reduction formula
$I_n=\int\frac{\mathrm{d}x}{(x^2+a^2)^n}$	$I_n=\frac{x}{2a^2(n-1)(x^2+a^2)^{n-1}}+\frac{2n-3}{2a^2(n-1)}I_{n-1}$
$I_{n,m}=\int\frac{\mathrm{d}x}{x^m(x^2+a^2)^n}$	$a^2I_{n,m}=I_{m,n-1}-I_{m-2,n}$
$I_{n,m}=\int\frac{x^m}{(x^2+a^2)^n}\mathrm{d}x$	$I_{n,m}=I_{m-2,n-1}-a^2I_{m-2,n}$

Integral	Reduction formula
$I_n=\int\frac{\mathrm{d}x}{(x^2-a^2)^n}$	$I_n=-\frac{x}{2a^2(n-1)(x^2-a^2)^{n-1}}-\frac{2n-3}{2a^2(n-1)}I_{n-1}$

$I_{n,m}=\int\frac{dx}{x^m(x^2-a^2)^n}$	$a^2I_{n,m}=I_{m-2,n}-I_{m,n-1}$
$I_{n,m}=\int\frac{x^m}{(x^2-a^2)^n}dx$	$I_{n,m}=I_{m-2,n-1}+a^2I_{m-2,n}$

Integral	Reduction formula
$I_n=\int\frac{dx}{(a^2-x^2)^n}$	$I_n=\frac{x}{2a^2(n-1)(a^2-x^2)^{n-1}}+\frac{2n-3}{2a^2(n-1)}I_{n-1}$
$I_{n,m}=\int\frac{dx}{x^m(a^2-x^2)^n}$	$a^2I_{n,m}=I_{m,n-1}+I_{m-2,n}$
$I_{n,m}=\int\frac{x^m}{(a^2-x^2)^n}dx$	$I_{n,m}=a^2I_{m-2,n}-I_{m-2,n-1}$

Integral	Reduction formula
$I_n=\int\frac{dx}{x^n(ax^2+bx+c)}$	$-cI_n=\frac{1}{x^{n-1}(n-1)}+bI_{n-1}+aI_{n-2}$
$I_{m,n}=\int\frac{x^m\,dx}{(ax^2+bx+c)^n}$	$I_{m,n}=-\frac{x^{m-1}}{a(2n-m-1)(ax^2+bx+c)^{n-1}}-\frac{b(n-m)}{a(2n-m-1)}I_{m-1,n}+\frac{c(m-1)}{a(2n-m-1)}I_{m-2,n}$
$I_{m,n}=\int\frac{dx}{x^m(ax^2+bx+c)^n}$	$-c(m-1)I_{m,n}=\frac{1}{x^{m-1}(ax^2+bx+c)^{n-1}}+a(m+2n-3)I_{m-2,n}+b(m+n-2)I_{m-1,n}$

Integral	Reduction formula
$I_1=\int(ax^2+bx+c)^n\,dx$	$8a(n+1)I_{n+\frac{1}{2}}=2(2ax+b)(ax^2+bx+c)^{n+\frac{1}{2}}+(2n+1)(4ac-b^2)I_{n-\frac{1}{2}}$
$I_2=\int\frac{1}{(ax^2+bx+c)^n}dx$	$(2n-1)(4ac-b^2)I_{n+\frac{1}{2}}=\frac{2(2ax+b)}{(ax^2+bx+c)^{n-\frac{1}{2}}}+8a(n-1)I_{n-\frac{1}{2}}$

note that by the laws of indices:

$$I_{n-\frac{1}{2}}=I_{\frac{2n+1}{2}}=\int\frac{1}{(ax^2+bx+c)^{\frac{2n+1}{2}}}dx=\int\frac{1}{\sqrt{(ax^2+bx+c)^{2n+1}}}dx$$

Transcendental Functions

The following integrals contain:

- Factors of sine.
- Factors of cosine.
- Factors of sine and cosine products and quotients.
- Products/quotients of exponential factors and powers of x.
- Products of exponential and sine/cosine factors.

Integral	Reduction formula
$I_n = \int x^n \sin ax\mathrm{d}x$	$a^2 I_n = -ax^n \cos ax + nx^{n-1} \sin ax - n(n-1)I_{n-2}$
$J_n = \int x^n \cos ax\mathrm{d}x$	$a^2 J_n = ax^n \sin ax + nx^{n-1} \cos ax - n(n-1)J_{n-2}$
$I_n = \int \frac{\sin ax}{x^n}\mathrm{d}x$ $J_n = \int \frac{\cos ax}{x^n}\mathrm{d}x$	$I_n = -\frac{\sin ax}{(n-1)x^{n-1}} + \frac{a}{n-1}J_{n-1}$ $J_n = -\frac{\cos ax}{(n-1)x^{n-1}} - \frac{a}{n-1}I_{n-1}$ the formulae can be combined to obtain separate equations in I_n: $J_{n-1} = -\frac{\cos ax}{(n-2)x^{n-2}} - \frac{a}{n-2}I_{n-2}$ $I_n = -\frac{\sin ax}{(n-1)x^{n-1}} - \frac{a}{n-1}\left[\frac{\cos ax}{(n-2)x^{n-2}} + \frac{a}{n-2}I_{n-2}\right]$ $\therefore I_n = -\frac{\sin ax}{(n-1)x^{n-1}} - \frac{a}{(n-1)(n-2)}\left(\frac{\cos ax}{x^{n-2}} + aI_{n-2}\right)$ and J_n: $I_{n-1} = -\frac{\sin ax}{(n-2)x^{n-2}} + \frac{a}{n-2}J_{n-2}$ $J_n = -\frac{\cos ax}{(n-1)x^{n-1}} - \frac{a}{n-1}\left[-\frac{\sin ax}{(n-2)x^{n-2}} + \frac{a}{n-2}J_{n-2}\right]$ $\therefore J_n = -\frac{\cos ax}{(n-1)x^{n-1}} - \frac{a}{(n-1)(n-2)}\left(-\frac{\sin ax}{x^{n-2}} + aJ_{n-2}\right)$
$I_n = \int \sin^n ax\mathrm{d}x$	$anI_n = -\sin^{n-1} ax \cos ax + a(n-1)I_{n-2}$
$J_n = \int \cos^n ax\mathrm{d}x$	$anJ_n = \sin ax \cos^{n-1} ax + a(n-1)J_{n-2}$

$I_n = \int \frac{\mathrm{d}x}{\sin^n ax}$	$(n-1)I_n = -\frac{\cos ax}{a\sin^{n-1} ax} + (n-2)I_{n-2}$
$J_n = \int \frac{\mathrm{d}x}{\cos^n ax}$	$(n-1)J_n = \frac{\sin ax}{a\cos^{n-1} ax} + (n-2)I_{n-2}$

Integral	Reduction formula
$I_{m,n} = \int \sin^m ax \cos^n ax \mathrm{d}x$	$I_{m,n} = \begin{cases} -\frac{\sin^{m-1} ax \cos^{n+1} ax}{a(m+n)} + \frac{m-1}{m+n} I_{m-2,n} \\ \frac{\sin^{m+1} ax \cos^{n-1} ax}{a(m+n)} + \frac{n-1}{m+n} I_{m,n-2} \end{cases}$
$I_{m,n} = \int \frac{\mathrm{d}x}{\sin^m ax \cos^n ax}$	$I_{m,n} = \begin{cases} \frac{1}{a(n-1)\sin^{m-1} ax \cos^{n-1} ax} + \frac{m+n-2}{n-1} I_{m,n-2} \\ -\frac{1}{a(m-1)\sin^{m-1} ax \cos^{n-1} ax} + \frac{m+n-2}{m-1} I_{m-2,n} \end{cases}$
$I_{m,n} = \int \frac{\sin^m ax}{\cos^n ax} \mathrm{d}x$	$I_{m,n} = \begin{cases} \frac{\sin^{m-1} ax}{a(n-1)\cos^{n-1} ax} - \frac{m-1}{n-1} I_{m-2,n-2} \\ \frac{\sin^{m+1} ax}{a(n-1)\cos^{n-1} ax} - \frac{m-n+2}{n-1} I_{m,n-2} \\ \frac{\sin^{m-1} ax}{a(m-n)\cos^{n-1} ax} + \frac{m-1}{m-n} I_{m-2,n} \end{cases}$
$I_{m,n} = \int \frac{\cos^m ax}{\sin^n ax} \mathrm{d}x$	$I_{m,n} = \begin{cases} -\frac{\cos^{m-1} ax}{a(n-1)\sin^{n-1} ax} - \frac{m-1}{n-1} I_{m-2,n-2} \\ -\frac{\cos^{m+1} ax}{a(n-1)\sin^{n-1} ax} - \frac{m-n+2}{n-1} I_{m,n-2} \\ \frac{\cos^{m-1} ax}{a(m-n)\sin^{n-1} ax} + \frac{m-1}{m-n} I_{m-2,n} \end{cases}$

Integral	Reduction formula
$I_n = \int x^n e^{ax} \mathrm{d}x$ $n > 0$	$I_n = \frac{x^n e^{ax}}{a} - \frac{n}{a} I_{n-1}$

$I_n = \int x^{-n} e^{ax}\,dx$ $n > 0$ $n \neq 1$	$I_n = \frac{-e^{ax}}{(n-1)x^{n-1}} + \frac{a}{n-1} I_{n-1}$
$I_n = \int e^{ax} \sin^n bx\,dx$	$I_n = \frac{e^{ax} \sin^{n-1} bx}{a^2 + (bn)^2}\left(a \sin bx - bn \cos bx\right) + \frac{n(n-1)b^2}{a^2 + (bn)^2} I_{n-2}$
$I_n = \int e^{ax} \cos^n bx\,dx$	$I_n = \frac{e^{ax} \cos^{n-1} bx}{a^2 + (bn)^2}\left(a \cos bx + bn \sin bx\right) + \frac{n(n-1)b^2}{a^2 + (bn)^2} I_{n-2}$

References

- Elena Irodionovna Obolashvili (2003). Higher order partial differential equations in Clifford analysis: effective solutions to problems. Birkhäuser. p. 101. ISBN 0-8176-4286-2
- Frank Ayres, Elliott Mendelson. Schaum's Outlines: Calculus. McGraw-Hill Professional 2008, ISBN 978-0-07-150861-2. pp. 244–248 (online copy, p. 244, at Google Books. Retrieved01, May 2020)
- Saff, Edward B.; Snider, Arthur David (2003). "Chapter 4". Fundamentals of Complex Analysis with Applications to Engineering, Science, and Mathematics (3rd ed.). ISBN 0-1390-7874-6
- Mathematical methods for physics and engineering, K.F. Riley, M.P. Hobson, S.J. Bence, Cambridge University Press, 2010, ISBN 978-0-521-86153-3
- Richard Courant & Fritz John (2000). Introduction to Calculus and Analysis: Vol. II/1, II/2. Classics in mathematics. Springer. p. 897. ISBN 3-540-66569-2
- M. Aslam Chaudhry & Syed M. Zubair (2001). On a Class of Incomplete Gamma Functions with Applications. CRC Press. p. Appendix C. ISBN 1-58488-143-7
- Thomas, George B.; Weir, Maurice D.; Hass, Joel (2010). Thomas' Calculus: Early Transcendentals (12th ed.). Addison-Wesley. ISBN 0-321-58876-2

6

Vectors and Vector Calculus

The integration and differentiation of vector fields is termed as vector calculus. There are various theorems which are studied within this field such as Stokes' theorem, Green's theorem and divergence theorem. The topics elaborated in this chapter will help in gaining a better perspective about these aspects of vectors and vector calculus.

Vector calculus (or vector analysis) is a branch of mathematics concerned with differentiation and integration of vector fields. The term "vector calculus" is sometimes used as a synonym for the broader subject of multivariable calculus, which includes vector calculus as well as partial differentiation and multiple integration. Vector calculus plays an important role in differential geometry and in the study of partial differential equations. It is used extensively in physics, especially in the description of electromagnetic and gravitational fields.

Basic Concepts

Fields A field is a continuous function that returns a number (or sets of numbers) for every point in space and time (x, t). There are three basic flavours of fields we will deal with scalar fields A scalar field f(x, y, z, t) returns a single number for every point in space and time. Examples include temperature, salinity, porosity, density. . . . vector fields A vector field F(x, y, z, t) returns a vector for every point in space and is readily visualized as a field of arrows. Examples include velocity, elastic displacements, electric or magnetic fields. tensor fields A second rank tensor field D(x, y, z, t) can be visualized as a field of ellipsoids (3 orthogonal vectors for every point). Examples include stress, strain, strain-rate.

Notation There are many different notations for scalars, vectors and tensors (and they're often mixed and matched); however, a few of the common ones are scalars scalars are usually shown in math italics e.g. f, g, t. . . vectors come in more flavours. when typeset they're usually bold-roman characters e.g. V, or the unit vectors i, j, k. When hand written they usually have a line underneath them. Vectors can also be written in component form as $\mathrm{v} = v_x \mathrm{i} + v_y \mathrm{j} + v_z \mathrm{k}$ or in index notation $\mathrm{v} = v_i \hat{e}_i$ where $\hat{e}_i$ is another representation of the unit vectors.

Tensors (actually second rank tensors) Typeset in sans-serif font D (or often just a bold), handwritten with two underbars, or by component D_{ij}. 2nd rank tensors are also conveniently represented by matrices.

Definitions of Basic Operations

Vector Dot Product

$$a \cdot b = a_i b_i = |a||b| \cos\theta$$

is a scalar that records the amount of vector a that lies in the direction of vector b (and vice versa). θ is the smallest angle between the two vectors.

Vector cross product c = a × b is a vector that is perpendicular to the plane spanned by vectors a and b. The direction that c points in is determined by the right hand rule. Note a × b = −b × a. The cross product is most easily calculated as the determinant of the matrix,

$$c = \begin{vmatrix} i & j & k \\ a_x & a_y & a_z \\ b_x & b_y & b_z \end{vmatrix}$$

Or,

$$c = (a_y b_z - a_z b_y)i - (a_x b_z - a_z b_x)j + (a_x b_y - a_y b_x)k$$

or in index notation as $c_i = \epsilon_{ijk} a_j b_k$ where ϵ_{ijk} is the horrid permutation symbol.

Tensor vector dot product is a vector formed by matrix multiplication of a tensor and a vector c = D · a. In the case of stress, the force acting on a plane with normal vector n is simply f = σ · n. Each component of the vector is most easily calculated in index notation with $c_i = D_{ij} a_j$ with summation implied over repeated indices (i.e. $c_1 = D_{11}a_1 + D_{12}a_2 + D_{13}a_3$ and so on for $i = 2,3$.

Vector Algebra

The algebraic (non-differential) operations in vector calculus are referred to as vector algebra, being defined for a vector space and then globally applied to a vector field. The basic algebraic operations consist of:

Operation	Notation	Description
Vector addition	$v_1 + v_2$	Addition of two vector fields, yielding a vector field.
Scalar multiplication	av	Multiplication of a scalar field and a vector field, yielding a vector field.
Dot product	$v_1 \cdot v_2$	Multiplication of two vector fields, yielding a scalar field.
Cross product	$v_1 \times v_2$	Multiplication of two vector fields in $\mathbb{R}^3$, yielding a (pseudo) vector field.

Also commonly used are the two triple products:

Operation	Notation	Description
Scalar triple product	$v_1 \cdot (v_2 \times v_3)$	The dot product of a vector and a cross product of two vectors.
Vector triple product	$v_1 \times (v_2 \times v_3)$	The cross product of a vector and a cross product of two vectors.

Operators and Theorems

Differential Operators

Vector calculus studies various differential operators defined on scalar or vector fields, which are typically expressed in terms of the del operator (∇), also known as "nabla". The three basic vector operators are:

Operation	Notation	Description	Notational analogy	Domain/Range
Gradient	$\operatorname{grad}(f) = \nabla f$	Measures the rate and direction of change in a scalar field.	Scalar multiplication	Maps scalar fields to vector fields.
Divergence	$\operatorname{div}(\mathbf{F}) = \nabla \cdot \mathbf{F}$	Measures the scalar of a source or sink at a given point in a vector field.	Dot product	Maps vector fields to scalar fields.
Curl	$\operatorname{curl}(\mathbf{F}) = \nabla \times \mathbf{F}$	Measures the tendency to rotate about a point in a vector field in $\mathbb{R}^3$.	Cross product	Maps vector fields to (pseudo)vector fields.
f denotes a scalar field and F denotes a vector field				

Also commonly used are the two Laplace operators:

Operation	Notation	Description	Domain/Range
Laplacian	$\Delta f = \nabla^2 f = \nabla \cdot \nabla f$	Measures the difference between the value of the scalar field with its average on infinitesimal balls.	Maps between scalar fields.
Vector Laplacian	$\nabla^2\mathbf{F} = \nabla(\nabla \cdot \mathbf{F}) - \nabla \times (\nabla \times \mathbf{F})$ F F f	Measures the difference between the value of the vector field with its average on infinitesimal balls.	Maps between vector fields.
f denotes a scalar field and F denotes a vector field			

A quantity called the Jacobian matrix is useful for studying functions when both the domain and range of the function are multivariable, such as a change of variables during integration.

Integral Theorems

The three basic vector operators have corresponding theorems which generalize the fundamental theorem of calculus to higher dimensions.

Theorem	Statement	Description
Gradient theorem	$\int_{L\subset\mathbb{R}^n} \nabla\varphi\cdot d\mathbf{r} = \varphi(\mathbf{q})-\varphi(\mathbf{p}) \quad for \quad L = L[p\to q]$	The line integral of the gradient of a scalar field over a curve L is equal to the change in the scalar field between the endpoints p and q of the curve.
Divergence theorem	$\underbrace{\int\cdots\int_{V\subset\mathbb{R}^n}}_{n} (\nabla\cdot\mathrm{F})\,dV = \underbrace{\oint\cdots\oint_{\partial V}}_{n-1} \mathrm{F}\cdot d\mathrm{S}$	The integral of the divergence of a vector field over an n-dimensional solid V is equal to the flux of the vector field through the (n−1)-dimensional closed boundary surface of the solid.
Curl (Kelvin–Stokes') theorem	$\iint_{\Sigma\subset\mathbb{R}^3} (\nabla\times\mathrm{F})\cdot d\Sigma = \oint_{\partial\Sigma} \mathrm{F}\cdot dr$	The integral of the curl of a vector field over a surface Σ in $\mathbb{R}^3$ is equal to the circulation of the vector field around the closed curve bounding the surface.
φ denotes a scalar field and F denotes a vector field		

In two dimensions, the divergence and curl theorems reduce to the Green's theorem:

Theorem	Statement	Description
Green's theorem	$\iint_{A\subset\mathbb{R}^2}\left(\frac{\partial M}{\partial x}-\frac{\partial L}{\partial y}\right)dA = \oint_{\partial A}(Ldx + Mdy)$	The integral of the divergence (or curl) of a vector field over some region A in $\mathbb{R}^2$ equals the flux (or circulation) of the vector field over the closed curve bounding the region.
For divergence, $F=(M,-L)$. For curl, $F=(L,M,0)$. L and M are functions of (x, y).		

Applications

Linear Approximations

Linear approximations are used to replace complicated functions with linear functions that are almost the same. Given a differentiable function $f(x,y)$ with real values, one can approximate $f(x,y)$ for (x,y) close to (a,b) by the formula:

$$f(x,y)\approx f(a,b)+\frac{\partial f}{\partial x}(a,b)(x-a)+\frac{\partial f}{\partial y}(a,b)(y-b).$$

The right-hand side is the equation of the plane tangent to the graph of $z=f(x,y)$ at (a,b).

Optimization

For a continuously differentiable function of several real variables, a point *P* (that is a set of values for the input variables, which is viewed as a point in R^n) is critical if all of the partial derivatives of the function are zero at *P*, or, equivalently, if its gradient is zero. The critical values are the values of the function at the critical points.

If the function is smooth, or, at least twice continuously differentiable, a critical point may be either a local maximum, a local minimum or a saddle point. The different cases may be distinguished by considering the eigenvalues of the Hessian matrix of second derivatives.

By Fermat's theorem, all local maxima and minima of a differentiable function occur at critical points. Therefore, to find the local maxima and minima, it suffices, theoretically, to compute the zeros of the gradient and the eigenvalues of the Hessian matrix at these zeros.

Physics and Engineering

Vector calculus is particularly useful in studying:

- Center of mass.
- Field theory.
- Kinematics.
- Maxwell's equations.

Generalizations

Different 3-Manifolds

Vector calculus is initially defined for Euclidean 3-space, $\mathbb{R}^3$ which has additional structure beyond simply being a 3-dimensional real vector space, namely: a norm (giving a notion of length) defined via an inner product (the dot product), which in turn gives a notion of angle, and an orientation, which gives a notion of left-handed and right-handed. These structures give rise to a volume form, and also the cross product, which is used pervasively in vector calculus.

The gradient and divergence require only the inner product, while the curl and the cross product also requires the handedness of the coordinate system to be taken into account.

Vector calculus can be defined on other 3-dimensional real vector spaces if they have an inner product (or more generally a symmetric nondegenerate form) and an orientation; note that this is less data than an isomorphism to Euclidean space, as it does not require a set of coordinates (a frame of reference), which reflects the fact that vector calculus is invariant under rotations (the special orthogonal group SO(3)).

More generally, vector calculus can be defined on any 3-dimensional oriented Riemannian manifold, or more generally pseudo-Riemannian manifold. This structure simply means that the tangent space at each point has an inner product (more generally, a symmetric nondegenerate form) and an orientation, or more globally that there is a symmetric nondegenerate metric tensor and an orientation, and works because vector calculus is defined in terms of tangent vectors at each point.

Other Dimensions

Most of the analytic results are easily understood, in a more general form, using the machinery of differential geometry, of which vector calculus forms a subset. Grad and div generalize immediately to other dimensions, as do the gradient theorem, divergence theorem, and Laplacian (yielding harmonic analysis), while curl and cross product do not generalize as directly.

From a general point of view, the various fields in (3-dimensional) vector calculus are uniformly seen as being k-vector fields: scalar fields are 0-vector fields, vector fields are 1-vector fields, pseudovector fields are 2-vector fields, and pseudoscalar fields are 3-vector fields. In higher dimensions there are additional types of fields (scalar/vector/pseudovector/pseudoscalar corresponding to $0/1/n-1/n$ dimensions, which is exhaustive in dimension 3), so one cannot only work with (pseudo)scalars and (pseudo)vectors.

In any dimension, assuming a nondegenerate form, grad of a scalar function is a vector field, and div of a vector field is a scalar function, but only in dimension 3 or 7 (and, trivially, in dimension 0 or 1) is the curl of a vector field a vector field, and only in 3 or 7 dimensions can a cross product be defined (generalizations in other dimensionalities either require $n-1$ vectors to yield 1 vector, or are alternative Lie algebras, which are more general antisymmetric bilinear products). The generalization of grad and div, and how curl may be generalized is elaborated at Curl: Generalizations; in brief, the curl of a vector field is a bivector field, which may be interpreted as the special orthogonal Lie algebra of infinitesimal rotations; however, this cannot be identified with a vector field because the dimensions differ – there are 3 dimensions of rotations in 3 dimensions, but 6 dimensions of rotations in 4 dimensions (and more generally $\binom{n}{2}=\frac{1}{2}n(n-1)$ dimensions of rotations in n dimensions).

There are two important alternative generalizations of vector calculus. The first, geometric algebra, uses k-vector fields instead of vector fields (in 3 or fewer dimensions, every k-vector field can be identified with a scalar function or vector field, but this is not true in higher dimensions). This replaces the cross product, which is specific to 3 dimensions, taking in two vector fields and giving as output a vector field, with the exterior product, which exists in all dimensions and takes in two vector fields, giving as output a bivector (2-vector) field. This product yields Clifford algebras as the algebraic structure on vector spaces (with an orientation and nondegenerate form). Geometric algebra is mostly used in generalizations of physics and other applied fields to higher dimensions.

The second generalization uses differential forms (k-covector fields) instead of vector fields or k-vector fields, and is widely used in mathematics, particularly in differential geometry, geometric topology, and harmonic analysis, in particular yielding Hodge theory on oriented pseudo-Riemannian manifolds. From this point of view, grad, curl, and div correspond to the exterior derivative of 0-forms, 1-forms, and 2-forms, respectively, and the key theorems of vector calculus are all special cases of the general form of Stokes' theorem.

From the point of view of both of these generalizations, vector calculus implicitly identifies mathematically distinct objects, which makes the presentation simpler but the underlying mathematical structure and generalizations less clear. From the point of view of geometric algebra, vector calculus implicitly identifies k-vector fields with vector fields or scalar functions: 0-vectors and 3-vectors with scalars, 1-vectors and 2-vectors with vectors. From the point of view of differential forms, vector calculus implicitly identifies k-forms with scalar fields or vector fields: 0-forms and 3-forms with scalar fields, 1-forms and 2-forms with vector fields. Thus for example the curl naturally takes as input a vector field or 1-form, but naturally has as output a 2-vector field or 2-form (hence pseudovector field), which is then interpreted as a vector field, rather than directly taking a vector field to a vector field; this is reflected in the curl of a vector field in higher dimensions not having as output a vector field.

Vector Operator

Three vector calculus operations which find many applications in physics are:

1. The divergence of a vector function:

$$\nabla \cdot E = \frac{\partial E_x}{\partial x} + \frac{\partial E_y}{\partial y} + \frac{\partial E_z}{\partial z}$$

2. The curl of a vector function:

$$\nabla \times E = \left(\frac{\partial E_z}{\partial y} - \frac{\partial E_y}{\partial z} \right) i + \left(\frac{\partial E_x}{\partial z} - \frac{\partial E_z}{\partial x} \right) j + \left(\frac{\partial E_y}{\partial x} - \frac{\partial E_x}{\partial y} \right) k$$

3. The Gradient of a scalar function:

$$\nabla f = \frac{\partial f}{\partial x} i + \frac{\partial f}{\partial y} j + \frac{\partial f}{\partial z} k$$

These examples of vector calculus operations are expressed in Cartesian coordinates, but they can be expressed in terms of any orthogonal coordinate system, aiding in the solution of physical problems which have other than rectangular symmetries.

Del

Vector differential operator `del' is represented by a symbol $\Delta$. Its another name is `nabla'. For three dimensional case, it is defined as:

$$\nabla = \text{id/ dx+ jd/ dy+ kd/ dz}$$

where I, j, k are unit vectors in the directions of co-ordinate axis X, Y and Z respectively.

The del operator can be applied to:

- a scalar function:

  $\vec{\nabla} z(x,y)$ $= \overrightarrow{e_x} \frac{\partial z}{\partial x} + \overrightarrow{e_y} \frac{\partial z}{\partial y}$

This generates separate values for the two differentials at each point in *x,y* space: $= \overrightarrow{e_x u_x} + \overrightarrow{e_y u},$

which can be written as a vector: $= \vec{u}(x,y).$

This *vector function* has an *x* and a y component at each point in (*x,y*) space. The resulting vector function is called the *gradient* of the original scalar function. Thus the electric field is the gradient of the electric potential.

- a vector function:

  $\vec{\nabla} \vec{u}(x,y)$ $= \overrightarrow{e_x} \frac{\partial \vec{u}}{\partial x} + \overrightarrow{e_y} \frac{\partial \vec{u}}{\partial y}$

The differentials apply separately to each component: $= \vec{e}_x \frac{\partial}{\partial x}\left(\vec{e}_x u_x + \vec{e}_y u_y\right) + \vec{e}_y \frac{\partial}{\partial y}\left(\vec{e}_x u_x + \vec{e}_y u_y\right).$

The unit vectors are orthogonal: $= \frac{\partial u_x}{\partial x} + \frac{\partial u_y}{\partial y},$

resulting in a scalar function: $= f(x,y)$

This scalar function $f(x,y)$ is known as the *divergence* of the original vector function.

Hence, the del operator turns scalar functions into vector functions and vector functions into scalar functions. If it is applied twice to a scalar function, the end result is another scalar function:

$$\nabla^2 z(x,y) = \vec{\nabla}\left(\vec{\nabla} z(x,y)\right) = f\,(x,y)$$

This is the 2nd derivative of $z(x,y)$ *along the steepest slope.*

The Del Operator in Polar Coordinates

In Cartesian coordinates, the del operator takes the same form when applied to scalar and vector functions. This is not the case in polar (spherical or cylindrical) coordinates.

In Cartesian coordinates the three-dimensional del operator is:

$$\nabla = \mathrm{i}\frac{\partial}{\partial x} + \mathrm{j}\frac{\partial}{\partial y} + \mathrm{k}\frac{\partial}{z},$$

and the horizontal component is:

$$\nabla_H = \mathrm{i}\frac{\partial}{\partial x} + \mathrm{j}\frac{\partial}{\partial y}.$$

Expressions for ∇ in various systems of curvilinear coordinates may be found in any textbook of vector analysis. In meteorology it is often convenient to use a thermodynamic function of state, such as pressure or potential temperature, as the vertical coordinate. If the chosen function is σ, then,

$$\nabla = \mathrm{i}\left(\frac{\partial}{\partial x}\right)_\sigma + \mathrm{j}\left(\frac{\partial}{\partial y}\right)_\sigma + \mathrm{k}\left(\frac{\partial \sigma}{\partial z}\right)\frac{\partial}{\partial \sigma},$$

where differentiation with respect to x and y is understood as carried out on surfaces of constant σ (the subscript usually being omitted). The horizontal component is now,

$$\nabla_\sigma = \mathrm{i}\frac{\partial}{\partial x} + \mathrm{j}\frac{\partial}{\partial y}.$$

If the quasi-hydrostatic approximation is justified, as in most meteorological contexts, pressure is a useful coordinate, and

$$\nabla = \nabla_p + \nabla_z p\left(\frac{\partial}{\partial p}\right) - \mathrm{k}\, pg\left(\frac{\partial}{\partial p}\right),$$

where g is the acceleration of gravity and ρ is the density. Here,

$$\nabla_p = \mathrm{i}\frac{\partial}{\partial x} + \mathrm{j}\frac{\partial}{\partial y},$$

with differentiation carried out in isobaric surfaces, and ∇_z is the horizontal gradient operator in Cartesian-altitude coordinates.

Directional Derivative

The directional derivative of a scalar field $f(x, y, z)$ in the direction $\vec{a}(x,y,z) = a_x\vec{e}_x + a_y\vec{e}_y + a_z\vec{e}_z$ is defined as:

$$\vec{a} \cdot \mathrm{grad}\, f = a_x\frac{\partial f}{\partial x} + a_y\frac{\partial f}{\partial y} + a_z\frac{\partial f}{\partial z} = \vec{a} \cdot (\nabla f)$$

This gives the rate of change of a field f in the direction of $\vec{a}$. In operator notation, the element in parentheses can be considered a single coherent unit; fluid dynamics uses this convention extensively, terming it the convective derivative—the "moving" derivative of the fluid.

Note that $(\vec{a} \cdot \nabla)$ is an operator that takes scalar to a scalar. It can be extended to operate on a vector, by separately operate on each of its components.

Laplacian

The Laplace operator is a scalar operator that can be applied to either vector or scalar fields; for cartesian coordinate systems it is defined as:

$$\Delta = \frac{\partial^2}{\partial x^2} + \frac{\partial^2}{\partial y^2} + \frac{\partial^2}{\partial z^2} = \nabla \cdot \nabla = \nabla$$

and the definition for more general coordinate systems is given in vector Laplacian.

The Laplacian is ubiquitous throughout modern mathematical physics, appearing for example in Laplace's equation, Poisson's equation, the heat equation, the wave equation, and the Schrödinger equation.

Tensor Derivative

Del can also be applied to a vector field with the result being a tensor. The tensor derivative of a vector field $\vec{v}$ (in three dimensions) is a 9-term second-rank tensor – that is, a 3×3 matrix – but can be denoted simply as $\nabla \otimes \vec{v}$, where $\otimes$ represents the dyadic product. This quantity is equivalent to the transpose of the Jacobian matrix of the vector field with respect to space. The divergence of the vector field can then be expressed as the trace of this matrix.

For a small displacement $\delta\vec{r}$, the change in the vector field is given by:

$$\delta\vec{v} = (\nabla \otimes \vec{v}) \cdot \delta\vec{r}$$

Product Rules

For vector calculus:

$$\nabla(fg) = f\nabla g + g\nabla f$$

$$\nabla(\vec{u}\cdot\vec{v}) = \vec{u}\times(\nabla\times\vec{v}) + \vec{v}\times(\nabla\times\vec{u}) + (\vec{u}\cdot\nabla)\vec{v} + (\vec{v}\cdot\nabla)\vec{u}$$

$$\nabla\cdot(f\vec{v}) = f(\nabla\cdot\vec{v}) + \vec{v}\cdot(\nabla f)$$

$$\nabla\cdot(\vec{u}\times\vec{v}) = \vec{v}\cdot(\nabla\times\vec{u}) - \vec{u}\cdot(\nabla\times\vec{v})$$

$$\nabla\times(f\vec{v}) = (\nabla f)\times\vec{v} + f(\nabla\times\vec{v})$$

$$\nabla\times(\vec{u}\times\vec{v}) = \vec{u}(\nabla\cdot\vec{v}) - \vec{v}(\nabla\cdot\vec{u}) + (\vec{v}\cdot\nabla)\vec{u} - (\vec{u}\cdot\nabla)\vec{v}$$

For matrix calculus (for which $\vec{u}\cdot\vec{v}$ can be written $\vec{u}^T\vec{v}$):

$$(\mathbf{A}\nabla)^T\vec{u} = \nabla^T(\mathbf{A}^T\vec{u}) - (\nabla^T\mathbf{A}^T)\vec{u}$$

Another relation of interest is the following, where $\vec{u}\otimes\vec{v}$ is the outer product tensor:

$$\nabla(\vec{u}\otimes\vec{v}) = (\nabla\cdot\vec{u})\vec{v} + (\vec{u}\cdot\nabla)\vec{v}$$

Precautions

Most of the above vector properties (except for those that rely explicitly on del's differential properties—for example, the product rule) rely only on symbol rearrangement, and must necessarily hold if the del symbol is replaced by any other vector. This is part of the value to be gained in notationally representing this operator as a vector.

Though one can often replace del with a vector and obtain a vector identity, making those identities mnemonic, the reverse is *not* necessarily reliable, because del does not commute in general.

A counterexample that relies on del's failure to commute:

$$(\vec{u}\cdot\vec{v})f \equiv (\vec{v}\cdot\vec{u})f$$

$$(\nabla\cdot\vec{v})f = \left(\frac{\partial v_x}{\partial x} + \frac{\partial v_y}{\partial y} + \frac{\partial v_z}{\partial z}\right)f = \frac{\partial v_x}{\partial x}f + \frac{\partial v_y}{\partial y}f + \frac{\partial v_z}{\partial z}f$$

$$(\vec{v}\cdot\nabla)f = \left(v_x\frac{\partial}{\partial x} + v_y\frac{\partial}{\partial y} + v_z\frac{\partial}{\partial z}\right)f = v_x\frac{\partial f}{\partial x} + v_y\frac{\partial f}{\partial y} + v_z\frac{\partial f}{\partial z}$$

$$\Rightarrow (\nabla\cdot\vec{v})f \neq (\vec{v}\cdot\nabla)f$$

A counterexample that relies on del's differential properties:

$$(\nabla x)\times(\nabla y) = \left(\vec{e}_x \frac{\partial x}{\partial x} + \vec{e}_y \frac{\partial x}{\partial y} + \vec{e}_z \frac{\partial x}{\partial z}\right) \times \left(\vec{e}_x \frac{\partial y}{\partial x} + \vec{e}_y \frac{\partial y}{\partial y} + \vec{e}_z \frac{\partial y}{\partial z}\right)$$

$$= (\vec{e}_x \cdot 1 + \vec{e}_y \cdot 0 + \vec{e}_z \cdot 0) \times (\vec{e}_x \cdot 0 + \vec{e}_y \cdot 1 + \vec{e}_z \cdot 0)$$

$$= \vec{e}_x \times \vec{e}_y$$

$$= \vec{e}_z$$

$$(\vec{u}x)\times(\vec{u}y) = xy(\vec{u}\times\vec{u})$$

$$= xy\vec{0}$$

$$= \vec{0}$$

Central to these distinctions is the fact that del is not simply a vector; it is a vector operator. Whereas a vector is an object with both a magnitude and direction, del has neither a magnitude nor a direction until it operates on a function.

For that reason, identities involving del must be derived with care, using both vector identities and *differentiation* identities such as the product rule.

Gradient

If $f(x, y)$ is a function of two variables x and y, then its gradient at (x, y) is the vector $[\partial f/\partial x, \partial f/\partial y]$. This notion can be defined also in one or three dimensions. In particular, if $f'(x)$ is a function of one variable then its gradient is the one-dimensional vector whose component is $f'(x)$, or in other words the derivative of f. The gradient in higher dimensions is a natural generalization of that, enabling us to calculate the way in which f changes in various directions just as $f'(x)$ measures the change of f in the only direction it could.

Note that the coordinates of the position of a point are written as (x, y), while coordinates of a vector, which measures relative position, are written $[x, y]$.

The gradient also has a natural intrinsic characterization, in that its components measure the rate of change of f in various directions, but in fact we can use it to measure rate of change in any direction. If $\Delta P = [\Delta x, \Delta y]$ is a small change from $P = (x, y)$ then we have approximately:

$$f(P+\Delta P) = f(p) + grad(f)\bullet\Delta P = f(x,y) + (\partial f + \partial x)\Delta x + (\partial f/\partial y)\Delta y.$$

In other words, the gradient bundles in one package (a vector field) the general idea of rate of change of f.

The Gradient of a Scalar Field

If $U(r) = U(x, y, z)$ is a scalar field, ie a scalar function of position $r = [x, y, z]$ in 3 dimensions, then its gradient at any point is defined in Cartesian co-ordinates by:

$$\text{grad}\,U = \frac{\partial U}{\partial x}\hat{i} + \frac{\partial U}{\partial y}\hat{j} + \frac{\partial U}{\partial z}\hat{k}$$

It is usual to define the vector operator which is called "del" or "nabla".

$$\nabla = \hat{\mathrm{i}}\frac{\partial}{\partial \mathrm{x}} + \hat{\mathrm{j}}\frac{\partial}{\partial y} + \hat{\mathrm{k}}\frac{\partial}{\partial z}$$

Then,

$$\mathrm{grad}\,U \equiv \nabla U$$

Worked examples of gradient evaluation,

1. $U = \mathrm{x}^2$

$$\Rightarrow \nabla U = \left(\frac{\partial}{\partial \mathrm{x}}\hat{\mathrm{i}} + \frac{\partial}{\partial y}\hat{\mathrm{j}} + \frac{\partial}{\partial z}\hat{\mathrm{k}}\right)\mathrm{x}^2 = 2\mathrm{x}\,\hat{\mathrm{i}}.$$

2. $U = \mathrm{r}^2$

$$\mathrm{r}^2 = x^2 + y^2 + z^2$$

$$\Rightarrow \nabla U = \left(\frac{\partial}{\partial \mathrm{x}}\hat{\mathrm{i}} + \frac{\partial}{\partial y}\hat{\mathrm{j}} + \frac{\partial}{\partial z}\hat{\mathrm{k}}\right)\left(x^2 + y^2 + z^2\right) = 2\mathrm{x}\,\hat{\mathrm{i}} + 2y\,\hat{\mathrm{j}} + 2z\,\hat{\mathrm{k}} = 2\mathrm{r}.$$

3. $U = c \cdot \mathrm{r}$, where c is constant.

$$\Rightarrow \nabla U = \left(\hat{\mathrm{i}}\frac{\partial}{\partial \mathrm{x}} + \hat{\mathrm{j}}\frac{\partial}{\partial y} + \hat{\mathrm{k}}\frac{\partial}{\partial z}\right)\left(c_1\mathrm{x} + c_2 y + c_3 z\right) = c_1\,\hat{\mathrm{i}} + c_2\,\hat{\mathrm{j}} + c_3\,\hat{\mathrm{k}} = c$$

4. $\mathrm{U} = \mathrm{U}(\mathrm{r})$, where $\mathrm{r} = \sqrt{\left(x^2 + y^2 + z^2\right)}$. NB NOT U(r).

U is a function of r alone so dU/dr exists. As U = U(x, y, z) also,

$$\frac{\partial U}{\partial \mathrm{x}} = \frac{dU}{dr}\frac{\partial r}{\partial \mathrm{x}} \qquad \frac{\partial U}{\partial y} = \frac{dU}{dr}\frac{\partial r}{\partial y} \qquad \frac{\partial U}{\partial z} = \frac{dU}{dr}\frac{\partial r}{\partial z}.$$

$$\Rightarrow \nabla U = \frac{\partial U}{\partial \mathrm{x}}\hat{\mathrm{i}} + \frac{\partial U}{\partial y}\hat{\mathrm{j}} + \frac{\partial U}{\partial z}\hat{\mathrm{k}} \quad = \frac{dU}{dr}\left(\frac{\partial r}{\partial \mathrm{x}}\hat{\mathrm{i}} + \frac{\partial r}{\partial y}\hat{\mathrm{j}} + \frac{\partial r}{\partial z}\hat{\mathrm{k}}\right)$$

But $\mathrm{r} = \sqrt{\left(x^2 + y^2 + z^2\right)}$, so $\partial r / \partial \mathrm{x} = \mathrm{x} / r$ and similarly for y, z.

$$\Rightarrow \nabla U = \frac{dU}{dr}\left(\frac{\mathrm{x}\,\hat{\mathrm{i}} + y\,\hat{\mathrm{j}} + z\,\hat{\mathrm{k}}}{r}\right) = \frac{dU}{dr}\left(\frac{\mathrm{r}}{r}\right).$$

Cylindrical and Spherical Coordinates

In cylindrical coordinates with a Euclidean metric, the gradient is given by:

$$\nabla f(\rho, \varphi, z) = \frac{\partial f}{\partial \rho}\mathbf{e}_\rho + \frac{1}{\rho}\frac{\partial f}{\partial \varphi}\mathbf{e}_\varphi + \frac{\partial f}{\partial z}\mathbf{e}_z$$

where ρ is the axial distance, φ is the azimuthal or azimuth angle, z is the axial coordinate, and e_ρ, e_φ and e_z are unit vectors pointing along the coordinate directions.

In spherical coordinates, the gradient is given by:

$$\nabla f(r,\theta,\varphi) = \frac{\partial f}{\partial r}\mathbf{e}_r + \frac{1}{r}\frac{\partial f}{\partial \theta}\mathbf{e}_\theta + \frac{1}{r\sin\theta}\frac{\partial f}{\partial \varphi}\mathbf{e}_\varphi$$

where r is the radial distance, φ is the azimuthal angle and θ is the polar angle, and $\mathbf{e}_r$, $\mathbf{e}_\theta$ and $\mathbf{e}_\varphi$ are again local unit vectors pointing in the coordinate directions (i.e. the normalized covariant basis).

General Coordinates

Using Einstein notation we can consider the gradient in general coordinates, which we write as $x^1, \ldots, x^i, \ldots, x^n$, where n is the number of dimensions of the domain. Here, the upper index refers to the number of the coordinate or component, so x^2 refers to the second component, and not the quantity x squared. The index variable i is used to refer to an arbitrary element, such as x^i. The gradient can then be written as:

$$\nabla f = \frac{\partial f}{\partial x^i}\mathbf{e}^i = \frac{\partial f}{\partial x^i} g^{ij}\mathbf{e}_j,$$

where $\mathbf{e}^i$ and $\mathbf{e}_j\left(=\frac{\partial \mathbf{x}}{\partial x^j}\right)$ refer to the unnormalized local contravariant and covariant bases respectively, g^{ij} is the inverse metric tensor, and the Einstein summation convention implies summation over i and j. In terms of the normalized bases, which we refer to as $\hat{\mathbf{e}}^i$ and $\hat{\mathbf{e}}_j$, using the identities $\|\mathbf{e}^i\| = \sqrt{g^{ii}}$ and $\|\mathbf{e}_j\| = \sqrt{g_{jj}}$, we can write:

$$\nabla f = \frac{\partial f}{\partial x^i}\sqrt{g^{ii}}\,\hat{\mathbf{e}}^i = \frac{\partial f}{\partial x^i} g^{ij}\sqrt{g_{jj}}\,\hat{\mathbf{e}}_j,$$

which will evaluate to the expressions given above for cylindrical and spherical coordinates.

Gradient and the Derivative or Differential

Linear Approximation to a Function

The gradient of a function f from the Euclidean space $\mathbb{R}^n$ to $\mathbb{R}$ at any particular point x_0 in $\mathbb{R}^n$ characterizes the best linear approximation to f at x_0. The approximation is as follows:

$$f(x) \approx f(x_0) + (\nabla f)_{x_0} \cdot (x - x_0)$$

for x close to x_0, where $(\nabla f)_{x0}$ is the gradient of f computed at x_0, and the dot denotes the dot product on $\mathbb{R}^n$. This equation is equivalent to the first two terms in the multivariable Taylor series expansion of f at x_0.

Differential or (Exterior) Derivative

The best linear approximation to a differentiable function:

$$f:\mathbb{R}^n \to \mathbb{R}$$

at a point x in $\mathbb{R}^n$ is a linear map from $\mathbb{R}^n$ to $\mathbb{R}$ which is often denoted by df_x or $Df(x)$ and called the differential or (total) derivative of f at x. The gradient is therefore related to the differential by the formula:

$$(\nabla f)_x \cdot v = df_x(v)$$

for any $v \in \mathbb{R}^n$. The function df, which maps x to df_x, is called the differential or exterior derivative of f and is an example of a differential 1-form.

If $\mathbb{R}^n$ is viewed as the space of (dimension n) column vectors (of real numbers), then one can regard df as the row vector with components,

$$\left(\frac{\partial f}{\partial x_1},\ldots,\frac{\partial f}{\partial x_n}\right)$$

so that $df_x(v)$ is given by matrix multiplication. Assuming the standard Euclidean metric on $\mathbb{R}^n$, the gradient is then the corresponding column vector, i.e.,

$$(\nabla f)_i = df_i^{\mathsf{T}}.$$

Gradient as a Derivative

Let U be an open set in $\mathbb{R}^n$. If the function $f: U \to \mathbb{R}$ is differentiable, then the differential of f is the (Fréchet) derivative of f. Thus ∇f is a function from U to the space $\mathbb{R}$ such that,

$$\lim_{h\to 0}\frac{\| f(x+h)-f(x)-\nabla f(x)\cdot h\|}{\| h\|}=0$$

where $\cdot$ is the dot product.

As a consequence, the usual properties of the derivative hold for the gradient:

Linearity

The gradient is linear in the sense that if f and g are two real-valued functions differentiable at the point $a \in \mathbb{R}^n$, and α and β are two constants, then $\alpha f + \beta g$ is differentiable at a, and moreover:

$$\nabla(\alpha f+\beta g)(a)=\alpha\nabla f(a)+\beta\nabla g(a).$$

Product Rule

If f and g are real-valued functions differentiable at a point $a \in \mathbb{R}^n$, then the product rule asserts that the product fg is differentiable at a, and,

$$\nabla(fg)(a)=f(a)\nabla g(a)+g(a)\nabla f(a).$$

Chain Rule

Suppose that $f: A \to \mathbb{R}$ is a real-valued function defined on a subset A of $\mathbb{R}^n$, and that f is differentiable at a point a. There are two forms of the chain rule applying to the gradient. First, suppose that the function g is a parametric curve; that is, a function $g : I \to \mathbb{R}^n$ maps a subset $I \subset \mathbb{R}$ into $\mathbb{R}^n$. If g is differentiable at a point $c \in I$ such that $g(c) = a$, then,

$$(f \circ g)'(c) = \nabla f(a) \cdot g'(c),$$

where $\circ$ is the composition operator: $(f \circ g)(x) = f(g(x))$.

More generally, if instead $I \subset \mathbb{R}^k$, then the following holds:

$$\nabla (f \circ g)(c) = (Dg(c))^{\mathrm{T}} (\nabla f(a))$$

where $(Dg)^{\mathrm{T}}$ denotes the transpose Jacobian matrix.

For the second form of the chain rule, suppose that $h : I \to \mathbb{R}$ is a real valued function on a subset I of $\mathbb{R}$, and that h is differentiable at the point $f(a) \in I$. Then,

$$\nabla (h \circ f)(a) = h'\big(f(a)\big)\nabla f(a).$$

Further Properties and Applications

Level Sets

A level surface, or isosurface, is the set of all points where some function has a given value.

If f is differentiable, then the dot product $(\nabla f)_x \cdot v$ of the gradient at a point x with a vector v gives the directional derivative of f at x in the direction v. It follows that in this case the gradient of f is orthogonal to the level sets of f. For example, a level surface in three-dimensional space is defined by an equation of the form $F(x, y, z) = c$. The gradient of F is then normal to the surface.

More generally, any embedded hypersurface in a Riemannian manifold can be cut out by an equation of the form $F(P) = 0$ such that $\mathrm{d}F$ is nowhere zero. The gradient of F is then normal to the hypersurface.

Similarly, an affine algebraic hypersurface may be defined by an equation $F(x_1,\ldots x_n) = 0$, where F is a polynomial. The gradient of F is zero at a singular point of the hypersurface (this is the definition of a singular point). At a non-singular point, it is a nonzero normal vector.

Conservative Vector Fields and the Gradient Theorem

The gradient of a function is called a gradient field. A (continuous) gradient field is always a conservative vector field: its line integral along any path depends only on the endpoints of the path, and can be evaluated by the gradient theorem (the fundamental theorem of calculus for line integrals). Conversely, a (continuous) conservative vector field is always the gradient of a function.

Generalizations

Gradient of a vector

Since the total derivative of a vector field is a linear mapping from vectors to vectors, it is a tensor quantity.

In rectangular coordinates, the gradient of a vector field $\mathbf{f} = (f^1, f^2, f^3)$ is defined by:

$$\nabla \mathbf{f} = g^{jk} \frac{\partial f^i}{\partial x^j} \mathbf{e}_i \otimes \mathbf{e}_k$$

where the Einstein summation notation is used and the tensor product of the vectors $\mathbf{e}_i$ and $\mathbf{e}_k$ is a dyadic tensor of type (2,0), or the transpose of the Jacobian matrix:

$$\frac{\partial f^i}{\partial x^j} = \frac{\partial (f^1, f^2, f^3)}{\partial (x^1, x^2, x^3)}.$$

In curvilinear coordinates, or more generally on a curved manifold, the gradient involves Christoffel symbols:

$$\nabla \mathbf{f} = g^{jk} \left(\frac{\partial f^i}{\partial x^j} + \Gamma^i{}_{jl} f^l \right) \mathbf{e}_i \otimes \mathbf{e}_k$$

where g^{jk} are the components of the inverse metric tensor and the $\mathbf{e}_i$ are the coordinate basis vectors.

Expressed more invariantly, the gradient of a vector field **f** can be defined by the Levi-Civita connection and metric tensor:

$$\nabla^a f^b = g^{ac} \nabla_c f^b,$$

where ∇_c is the connection.

Riemannian Manifolds

For any smooth function f on a Riemannian manifold (M, g), the gradient of f is the vector field ∇f such that for any vector field X,

$$g(\nabla f, X) = \partial_X f, \qquad i.e., \quad g_x((\nabla f)_x, X_x) = (\partial_X f)(x)$$

where $g_x(,)$ denotes the inner product of tangent vectors at x defined by the metric g and $\partial_X f$ is the function that takes any point $x \in M$ to the directional derivative of f in the direction X, evaluated at x. In other words, in a coordinate chart φ from an open subset of M to an open subset of $\mathbb{R}^n$, $(\partial_X f)(x)$ is given by:

$$\sum_{j=1}^{n} X^j(\varphi(x)) \frac{\partial}{\partial x_j} (f \circ \varphi^{-1}) \Big|_{\varphi(x)}$$

where X^j denotes the jth component of X in this coordinate chart.

So, the local form of the gradient takes the form:

$$\nabla f = g^{ik}\frac{\partial f}{\partial x^k}\mathbf{e}_i.$$

Generalizing the case $M = \mathbb{R}^n$, the gradient of a function is related to its exterior derivative, since,

$$(\partial_X f)(x) = (df)_x(X_x).$$

More precisely, the gradient ∇f is the vector field associated to the differential 1-form df using the musical isomorphism,

$$\sharp = \sharp^g : T^*M \to TM$$

(called "sharp") defined by the metric g. The relation between the exterior derivative and the gradient of a function on $\mathbb{R}^n$ is a special case of this in which the metric is the flat metric given by the dot product.

Divergence

The divergence computes a scalar quantity from a vector field by differentiation.

If a (x, y , z) is a vector function of position in 3 dimensions, that is $a = a_1\,\hat{\imath} + a_2\,\hat{\jmath} + a_3\,\hat{k}$, then its divergence at any point is defined in Cartesian co-ordinates by:

$$\text{div } a = \frac{\partial a_1}{a_x} + \frac{\partial a_2}{\partial_y} + \frac{\partial a_3}{\partial_z}$$

We can write this in a simplified notation using a scalar product with the Δ vector differential operator:

$$\text{div } a = \left(\hat{\imath}\frac{\partial}{\partial x} + \hat{\jmath}\frac{\partial}{\partial y} + \hat{k}\frac{\partial}{\partial z}\right)\cdot a = \nabla \cdot a$$

Notice that the divergence of a vector field is a scalar field.

Examples of divergence evaluation.

	a	div a
1)	$x\hat{\imath}$	1
2)	$r\left(= x\hat{\imath} + y\hat{\jmath} + z\hat{k}\right)$	3
3)	r/r^3	0
4)	rc, for c constant	(r . c)/r

The x component of r/r^3 is $x.\left(x^2+y^2+z^2\right)^{-3/2}$, and we need to find $\partial/\partial x$ of it.

$$\frac{\partial}{\partial x}x.\left(x^2+y^2+z^2\right)^{-3/2} = 1.\left(x^2+y^2+z^2\right)^{-3/2} + x\frac{-3}{2}\left(x^2+y^2+z^2\right)^{-5/2}.2x$$

$$= r^{-3}\left(1 - 3x^2 r^{-2}\right).$$

The terms in y and z are similar, so that.

$$\text{div}\left(\text{r}/\text{r}^3\right) = r^{-3}\left(3 - 3\left(x^2 + y^2 + z^2\right)r^{-2}\right) = r^{-3}(3-3)$$

$$= 0$$

The Significance of Divergence

Consider a typical vector field, water flow, and denote it by a(r). This vector has magnitude equal to the mass of water crossing a unit area perpendicular to the direction of a per unit time.

Now take an infinitesimal volume element dV and figure out the balance of the flow of a in and out of dV.

To be specific, consider the volume element dV = *dx dy dz* in Cartesian co-ordinates, and think first about the face of area dxdz perpendicular to the y axis and facing outwards in the negative y direction. (That is, the one with surface area $d\text{S} = -dx\, dz\hat{\text{j}}$.)

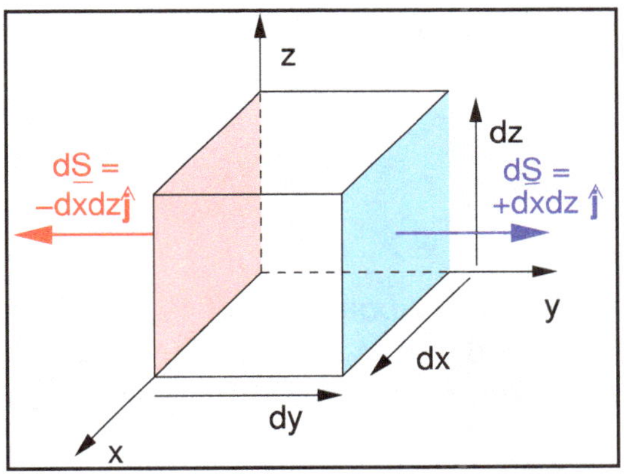

Elemental volume for calculating divergence.

The component of the vector a normal to this face is $\text{a}\cdot\hat{\text{j}} = \text{a}_y$, and is pointing inwards, and so the its contribution to the OUTWARD flux from this surface is:

$$\text{a}\cdot d\text{S} = -\text{a}_y(x, y, z)\,dzdx,$$

(By the way, flux here denotes mass per unit time.)

A similar contribution, but of opposite sign, will arise from the opposite face, but we must remember that we have moved along y by an amount dy, so that this OUTWARD amount is:

$$\text{a}_y\,(x, y+dy, z)\,dzdx = \left(\text{a}_y + \frac{\partial \text{a}_y}{\partial y}\right)dxdz$$

The total outward amount from these two faces is:

$$\frac{\partial \text{a}_y}{\partial y}dydxdz = \frac{\partial \text{a}_y}{\partial y}dV$$

Summing the other faces gives a total outward flux of:

$$\left(\frac{\partial a_x}{\partial x}+\frac{\partial a_y}{\partial y}+\frac{\partial a_z}{\partial z}\right)dV = \nabla \cdot a\, dV$$

The divergence of a vector field represents the flux generation per unit volume at each point of the field. (Divergence because it is an efflux not an influx.)

Physical Interpretation of Divergence

In physical terms, the divergence of a three-dimensional vector field is the extent to which the vector field flow behaves like a source at a given point. It is a local measure of its "outgoingness" – the extent to which there is more of some quantity exiting an infinitesimal region of space than entering it. If the divergence is nonzero at some point then there is compression or expansion at that point.

More rigorously, the divergence of a vector field F at a point p can be defined as the limit of the net flow of F across the smooth boundary of a three-dimensional region V divided by the volume of V as V shrinks to p. Formally,

$$\operatorname{div}\mathbf{F}\big|_p = \lim_{V\to\{p\}} \iint_{S(V)} \frac{\mathbf{F}\cdot\hat{\mathbf{n}}}{|V|}dS$$

where $|V|$ is the volume of V, $S(V)$ is the boundary of V, and the integral is a surface integral with $\hat{n}$ being the outward unit normal to that surface. The result, div F, is a function of p. From this definition it also becomes obvious that div F can be seen as the *source density* of the flux of F.

In light of the physical interpretation, a vector field with zero divergence everywhere is called *incompressible* or *solenoidal* – in which case any closed surface has no net flow across it.

The intuition that the sum of all sources minus the sum of all sinks should give the net flow outwards of a region is made precise by the divergence theorem.

Cartesian Coordinates

Let x, y, z be a system of Cartesian coordinates in 3-dimensional Euclidean space, and let i, j, k be the corresponding basis of unit vectors. The divergence of a continuously differentiable vector field F = Ui + Vj + Wk is defined as the scalar-valued function:

$$\operatorname{div}\mathbf{F} = \nabla\cdot\mathbf{F} = \left(\frac{\partial}{\partial x},\frac{\partial}{\partial y},\frac{\partial}{\partial z}\right)\cdot(U,V,W) = \frac{\partial U}{\partial x}+\frac{\partial V}{\partial y}+\frac{\partial W}{\partial z}.$$

Although expressed in terms of coordinates, the result is invariant under rotations, as the physical interpretation suggests. This is because the trace of the Jacobian matrix of an N-dimensional vector field F in N-dimensional space is invariant under any invertible linear transformation.

The common notation for the divergence $\nabla \cdot \mathbf{F}$ is a convenient mnemonic, where the dot denotes an operation reminiscent of the dot product: take the components of the ∇ operator, apply them to the corresponding components of F, and sum the results. Because applying an operator is different from multiplying the components, this is considered an abuse of notation.

The divergence of a continuously differentiable second-order tensor field ε is a first-order tensor field:

$$\overrightarrow{\operatorname{div}}(\varepsilon) = \begin{bmatrix} \frac{\partial \varepsilon_{xx}}{\partial x} + \frac{\partial \varepsilon_{yx}}{\partial y} + \frac{\partial \varepsilon_{zx}}{\partial z} \\ \frac{\partial \varepsilon_{xy}}{\partial x} + \frac{\partial \varepsilon_{yy}}{\partial y} + \frac{\partial \varepsilon_{zy}}{\partial z} \\ \frac{\partial \varepsilon_{xz}}{\partial x} + \frac{\partial \varepsilon_{yz}}{\partial y} + \frac{\partial \varepsilon_{zz}}{\partial z} \end{bmatrix}$$

Cylindrical Coordinates

For a vector expressed in local unit cylindrical coordinates as:

$$\mathbf{F} = \mathbf{e}_r F_r + \mathbf{e}_\theta F_\theta + \mathbf{e}_z F_z,$$

where e_a is the unit vector in direction a, the divergence is:

$$\operatorname{div} \mathbf{F} = \nabla \cdot \mathbf{F} = \frac{1}{r}\frac{\partial}{\partial r}(rF_r) + \frac{1}{r}\frac{\partial F_\theta}{\partial \theta} + \frac{\partial F_z}{\partial z}.$$

The use of local coordinates is vital for the validity of the expression. If we consider x the position vector and the functions $r(\mathbf{x})$, $\theta(\mathbf{x})$, and $z(\mathbf{x})$, which assign the corresponding global cylindrical coordinate to a vector, in general $r(\mathbf{F}(\mathbf{x})) \neq F_r(\mathbf{x})$, $\theta(\mathbf{F}(\mathbf{x})) \neq F_\theta(\mathbf{x})$, and $z(\mathbf{F}(\mathbf{x})) \neq F_z(\mathbf{x})$. In particular, if we consider the identity function $\mathbf{F}(\mathbf{x}) = \mathbf{x}$, we find that:

$$\theta(\mathbf{F}(\mathbf{x})) = \theta \neq F_\theta(\mathbf{x}) = 0.$$

Spherical Coordinates

In spherical coordinates, with θ the angle with the z axis and φ the rotation around the z axis, and $\mathbf{F}$ again written in local unit coordinates, the divergence is:

$$\operatorname{div} \mathbf{F} = \nabla \cdot \mathbf{F} = \frac{1}{r^2}\frac{\partial}{\partial r}(r^2 F_r) + \frac{1}{r\sin\theta}\frac{\partial}{\partial \theta}(\sin\theta F_\theta) + \frac{1}{r\sin\theta}\frac{\partial F_\varphi}{\partial \varphi}.$$

General Coordinates

Using Einstein notation we can consider the divergence in general coordinates, which we write as $x^1, \ldots, x^i, \ldots, x^n$, where n is the number of dimensions of the domain. Here, the upper index refers to the number of the coordinate or component, so x^2 refers to the second component, and not the quantity x squared. The index variable i is used to refer to an arbitrary element, such as x^i. The divergence can then be written via the Voss- Weyl formula, as:

$$\operatorname{div}(\mathbf{F}) = \frac{1}{\rho}\frac{\partial\left(\rho F^i\right)}{\partial x^i}$$

where ρ is the local coefficient of the volume element and F^i are the components of F with respect to the local unnormalized covariant basis (sometimes written as $\mathbf{e}_i = \frac{\partial \mathbf{x}}{\partial x^i}$). The Einstein notation implies summation over i, since it appears as both an upper and lower index.

The volume coefficient ρ is a function of position which depends on the coordinate system. In Cartesian, cylindrical and polar coordinates, $\rho = 1, r,$ and $r^2 \sin\theta$ respectively, using the same conventions as above. It can also be expressed as $\rho = \sqrt{\det g_{ab}}$, where g_{ab} is the metric tensor. Since the determinant is a scalar quantity which doesn't depend on the indices, we can suppress them and simply write $\rho = \sqrt{\det g}$. Another expression comes from computing the determinant of the Jacobian for transforming from Cartesian coordinates, which for n=3 gives $\rho = \left|\frac{\partial(x,y,z)}{\partial(x^1,x^2,x^3)}\right|$.

Some conventions expect all local basis elements to be normalized to unit length. If we write $\hat{\mathbf{e}}_i$ for the normalized basis, and $\hat{F}^i$ for the components of F with respect to it, we have that $\mathbf{F} = F^i\mathbf{e}_i = F^i \|\mathbf{e}_i\| \frac{\mathbf{e}_i}{\|\mathbf{e}_i\|} = F^i\sqrt{g_{ii}}\,\hat{\mathbf{e}}_i = \hat{F}^i\hat{\mathbf{e}}_i$, using one of the properties of the metric tensor. By dotting both sides of the last equality with the contravariant element $\hat{\mathbf{e}}^i$, we can conclude that $F^i = \frac{\hat{F}^i}{\sqrt{g_{ii}}}$. After substituting, the formula becomes:

$$\operatorname{div}(\mathbf{F}) = \frac{1}{\rho}\frac{\partial\left(\frac{\rho}{\sqrt{g_{ii}}}\hat{F}^i\right)}{\partial x^i} = \frac{1}{\sqrt{\det g}}\frac{\partial\left(\sqrt{\frac{\det g}{g_{ii}}}\hat{F}^i\right)}{\partial x^i}.$$

Decomposition Theorem

It can be shown that any stationary flux v(r) that is at least twice continuously differentiable in $\mathbb{R}^3$ and vanishes sufficiently fast for |r| → ∞ can be decomposed into an *irrotational part* E(r) and a *source-free part* B(r). Moreover, these parts are explicitly determined by the respective *source densities* and *circulation densities*:

For the irrotational part one has,

$$\mathbf{E} = -\nabla\Phi(\mathbf{r}),$$

with,

$$\Phi(\mathbf{r}) = \int_{\mathbb{R}^3} d^3\mathbf{r}'\,\frac{\operatorname{div}\mathbf{v}(\mathbf{r}')}{4\pi\left|\mathbf{r}-\mathbf{r}'\right|}.$$

The source-free part, B, can be similarly written: one only has to replace the *scalar potential* $\Phi(r)$ by a *vector potential* A(r) and the terms $-\nabla\Phi$ by $+\nabla \times A$, and the source density div v by the circulation density $\nabla \times v$.

This "decomposition theorem" is a by-product of the stationary case of electrodynamics. It is a special case of the more general Helmholtz decomposition which works in dimensions greater than three as well.

Properties

The following properties can all be derived from the ordinary differentiation rules of calculus. Most importantly, the divergence is a linear operator, i.e.

$$\operatorname{div}(a\mathbf{F} + b\mathbf{G}) = a \operatorname{div}\mathbf{F} + b \operatorname{div}\mathbf{G}$$

for all vector fields **F** and **G** and all real numbers *a* and *b*.

There is a product rule of the following type: if φ is a scalar-valued function and F is a vector field, then,

$$\operatorname{div}(\varphi\mathbf{F}) = \operatorname{grad}\varphi \cdot \mathbf{F} + \varphi \operatorname{div}\mathbf{F},$$

or in more suggestive notation,

$$\nabla \cdot (\varphi\mathbf{F}) = (\nabla\varphi) \cdot \mathbf{F} + \varphi(\nabla \cdot \mathbf{F}).$$

Another product rule for the cross product of two vector fields F and G in three dimensions involves the curl and reads as follows:

$$\operatorname{div}(\mathbf{F} \times \mathbf{G}) = \operatorname{curl}\mathbf{F} \cdot \mathbf{G} - \mathbf{F} \cdot \operatorname{curl}\mathbf{G},$$

or

$$\nabla \cdot (\mathbf{F} \times \mathbf{G}) = (\nabla \times \mathbf{F}) \cdot \mathbf{G} - \mathbf{F} \cdot (\nabla \times \mathbf{G}).$$

The Laplacian: Div (GradU) of a Scalar Field

Recall that gradU of any scalar field U is a vector field. Recall also that we can compute the divergence of any vector field. So we can certainly compute div (gradU), even if we don't know what it means yet.

Here is where the ∇ operator starts to be really handy.

$$\nabla.(\nabla U) = \left(\hat{i}\frac{\partial}{\partial x} + \hat{j}\frac{\partial}{\partial y} + \hat{k}\frac{\partial}{\partial z}\right) \cdot \left(\left(\hat{i}\frac{\partial}{\partial x} + \hat{j}\frac{\partial}{\partial y} + \hat{k}\frac{\partial}{\partial z}\right)U\right)$$

$$= \left(\left(\hat{i}\frac{\partial}{\partial x} + \hat{j}\frac{\partial}{\partial y} + \hat{k}\frac{\partial}{\partial z}\right) \cdot \left(\hat{i}\frac{\partial}{\partial x} + \hat{j}\frac{\partial}{\partial y} + \hat{k}\frac{\partial}{\partial z}\right)\right)U$$

$$= \left(\frac{\partial^2}{\partial x^2} + \frac{\partial^2}{\partial y^2} + \frac{\partial^2}{\partial z^2} \right) U$$

$$= \left(\frac{\partial^2 U}{\partial x^2} + \frac{\partial^2 U}{\partial y^2} + \frac{\partial^2 U}{\partial z^2} \right)$$

This last expression occurs frequently in engineering science (you will meet it next in solving Laplace's Equation in partial differential equations). For this reason, the operator ∇^2 is called the "Laplacian":

$$\nabla^2 U \left(\frac{\partial^2}{\partial x^2} + \frac{\partial^2}{\partial y^2} + \frac{\partial^2}{\partial z^2} \right) U$$

Laplace's equation itself is:

$$\nabla^2 U = 0$$

Examples of $\nabla^2 U$ evaluation:

U	$\nabla^2 U$
1) $r^2 \left(= x^2 + y^2 + z^2\right)$	6
2) $xy^2 z^3$	$2xz^3 + 6xy^2 z$
3) $1/r$	0

Let's prove example:

$$1/r = \left(x^2 + y^2 + z^2\right)^{-1/2}$$

$$\frac{\partial}{\partial x}\frac{\partial}{\partial x}\left(x^2 + y^2 + z^2\right)^{-1/2} = \frac{\partial}{\partial x} - x.\left(x^2 + y^2 + z^2\right)^{-3/2}$$

$$= -\left(x^2 + y^2 + z^2\right)^{-3/2} + 3x.x.\left(x^2 + y^2 + z^2\right)^{-5/2}$$

$$= \left(1/r^3\right)\left(-1 + 3x^2/r^2\right)$$

Adding up similar terms for y and z,

$$\nabla^2 \frac{1}{r} = \frac{1}{r^3}\left(-3 + 3\frac{\left(x^2 + y^2 + x^2\right)}{r^2} \right) = 0$$

Relation with the Exterior Derivative

One can express the divergence as a particular case of the exterior derivative, which takes a 2-form to a 3-form in $\mathbb{R}^3$. Define the current two-form as:

$$j = F_1 dy \wedge dz + F_2 dz \wedge dx + F_3 dx \wedge dy.$$

It measures the amount of "stuff" flowing through a surface per unit time in a "stuff fluid" of density $\rho = 1\, dx \wedge dy \wedge dz$ moving with local velocity F. Its exterior derivative dj is then given by:

$$dj = \left(\frac{\partial F_1}{\partial x} + \frac{\partial F_2}{\partial y} + \frac{\partial F_3}{\partial z} \right) dx \wedge dy \wedge dz = (\nabla \cdot \mathbf{F})\rho.$$

Thus, the divergence of the vector field F can be expressed as:

$$\nabla \cdot \mathbf{F} = \star d \star \left(\mathbf{F}^\flat \right).$$

Here the superscript $\flat$ is one of the two musical isomorphisms, and $\star$ is the Hodge dual. Working with the current two-form and the exterior derivative is usually easier than working with the vector field and divergence, because unlike the divergence, the exterior derivative commutes with a change of (curvilinear) coordinate system.

Generalizations

The divergence of a vector field can be defined in any number of dimensions. If,

$$\mathbf{F} = (F_1, F_2, \ldots F_n),$$

in a Euclidean coordinate system with coordinates $x_1, x_2, \ldots, x_n$, define,

$$\operatorname{div} \mathbf{F} = \nabla \cdot \mathbf{F} = \frac{\partial F_1}{\partial x_1} + \frac{\partial F_2}{\partial x_2} + \cdots + \frac{\partial F_n}{\partial x_n}.$$

The appropriate expression is more complicated in curvilinear coordinates.

In the case of one dimension, F reduces to a regular function, and the divergence reduces to the derivative.

For any n, the divergence is a linear operator, and it satisfies the "product rule":

$$\nabla \cdot (\varphi \mathbf{F}) = (\nabla \varphi) \cdot \mathbf{F} + \varphi (\nabla \cdot \mathbf{F})$$

for any scalar-valued function φ.

The divergence of a vector field extends naturally to any differentiable manifold of dimension n that has a volume form (or density) μ, e.g. a Riemannian or Lorentzian manifold. Generalising the construction of a two-form for a vector field on $\mathbb{R}^3$, on such a manifold a vector field X defines an $(n - 1)$-form $j = i_X \mu$ obtained by contracting X with μ. The divergence is then the function defined by,

$$dj = (\operatorname{div} X)\mu.$$

Standard formulas for the Lie derivative allow us to reformulate this as:

$$\mathcal{L}_X \mu = (\operatorname{div} X)\mu.$$

This means that the divergence measures the rate of expansion of a volume element as we let it flow with the vector field.

On a pseudo-Riemannian manifold, the divergence with respect to the metric volume form can be computed in terms of the Levi-Civita connection ∇:

$$\operatorname{div} X = \nabla \cdot X = X^{a}{}_{;a},$$

where the second expression is the contraction of the vector field valued 1- form ∇X with itself and the last expression is the traditional coordinate expression from Ricci calculus.

An equivalent expression without using connection is:

$$\operatorname{div}(X) = \frac{1}{\sqrt{\det g}} \partial_a \left(\sqrt{\det g}\, X^a \right),$$

where g is the metric and ∂_a denotes the partial derivative with respect to coordinate x^a.

Divergence can also be generalised to tensors. In Einstein notation, the divergence of a contravariant vector F^μ is given by:

$$\nabla \cdot \mathbf{F} = \nabla_\mu F^\mu,$$

where ∇_μ denotes the covariant derivative.

Equivalently, some authors define the divergence of a mixed tensor by using the musical isomorphism ♯: if T is a (p, q)-tensor (p for the contravariant vector and q for the covariant one), then we define the *divergence of T* to be the $(p, q-1)$-tensor:

$$(\operatorname{div} T)(Y_1,\ldots,Y_{q-1}) = \operatorname{trace}\big(X \mapsto \sharp(\nabla T)(X,\cdot,Y_1,\ldots,Y_{q-1})\big);$$

that is, we take the trace over the *first two* covariant indices of the covariant derivative.

Curl

The curl of a vector field A, denoted by curl A or $\nabla \times A$, is a vector whose magnitude is the maximum net circulation of A per unit area as the area tends to zero and whose direction is the normal direction of the area when the area is oriented to make the net circulation maximum.

Cross it with a vector field $\nabla \times \mathrm{a}$.

This gives the curl of a vector field.

$$\nabla \times \mathrm{a} \equiv \operatorname{curl}(\mathrm{a})$$

We can follow the pseudo-determinant recipe for vector products, so that,

$$\nabla \times \mathrm{a} = \begin{vmatrix} \hat{\mathrm{i}} & \hat{\mathrm{j}} & \hat{\mathrm{k}} \\ \dfrac{\partial}{\partial \mathrm{x}} & \dfrac{\partial}{\partial y} & \dfrac{\partial}{\partial z} \\ \partial \mathrm{x} & \partial y & \partial z \end{vmatrix}$$

$$= \left(\frac{\partial \mathrm{a}_z}{\partial y} - \frac{\partial \mathrm{a}_y}{\partial z} \right)\hat{\mathrm{i}} + \left(\frac{\partial \mathrm{a}_x}{\partial z} - \frac{\partial \mathrm{a}_z}{\partial \mathrm{x}} \right)\hat{\mathrm{j}} + \left(\frac{\partial \mathrm{a}_y}{\partial \mathrm{x}} - \frac{\partial \mathrm{a}_x}{\partial y} \right)\hat{\mathrm{k}}$$

Examples of Curl Evaluation

	a	$\nabla \times a$
1)	$-y\hat{i}+x\hat{j}$	$2\hat{k}$
2)	$x^2 y^2 \hat{k}$	$2x^2 y\hat{i}' - 2xy^2\hat{j}$

The Signficance of Curl

Perhaps the first example gives a clue. The field $a = -y\hat{i}+x\hat{j}$ is sketched in Figure. (It is the field you would calculate as the velocity field of an object rotating with $\grave{u} = [0,0,1]$.) This field has a curl of $2\hat{k}$, which is in the r-h screw sense out of the page. We can also see that a field like this must give a finite value to the line integral around the complete loop $\oint_C a \cdot dr$.

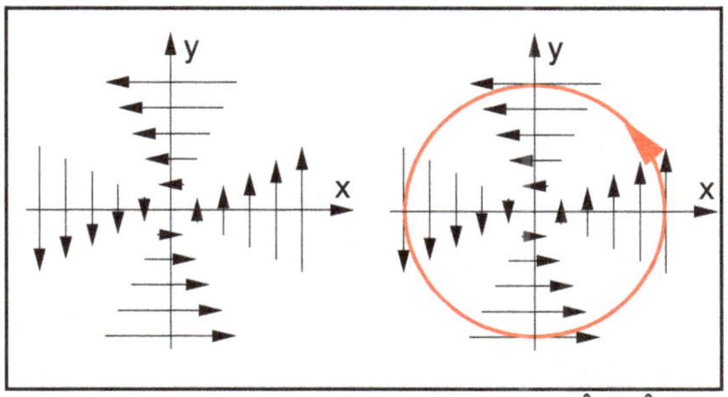

A rough sketch of the vector field $-y\hat{i}+x\hat{j}$.

In fact curl is closely related to the line integral around a loop.

The circulation of a vector a round any closed curve C is defined to be $\oint_C a \cdot dr$ and the curl of the vector field a represents the vorticity, or circulation per unit area, of the field.

Our proof uses the small rectangular element dx by dy shown in Figure.

Consider the circulation round the perimeter of a rectangular element.

The fields in the x direction at the bottom and top are:

$$a_x(x,y,z) \quad \text{and} \quad a_x(x,y+dy,z) = a_x(x,y,z) + \frac{\partial a_x}{\partial y} dy,$$

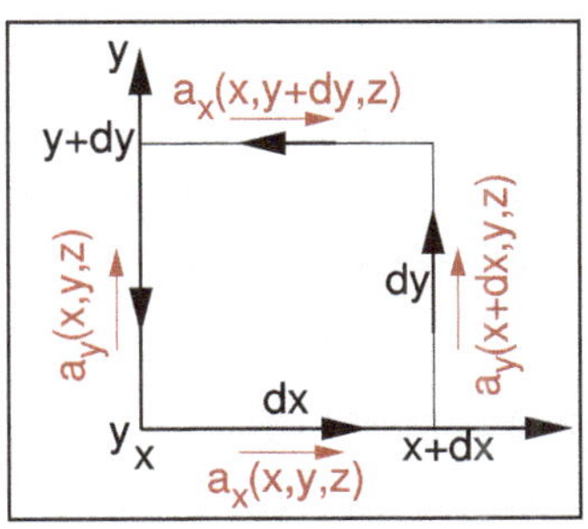

A small element used to calculate curl.

and the fields in the y direction at the left and right are,

$$a_y(x,y,z) \quad \text{and} \quad a_y(x+dx,y,z) = a_y(x,y,z) + \frac{\partial a_y}{\partial x}dx$$

Starting at the bottom and working round in the anticlockwise sense, the four contributions to the circulation dC are therefore as follows, where the minus signs take account of the path being opposed to the field:

$$\begin{aligned} dC &= +[a_x\, dx] + [a_y(x+dx,y,z)dy] - [a_x(x,y+dy,z)dx] - [ay\, dy] \\ &= +[a_x\, dx] + \left[\left(a_y + \frac{\partial a_y}{\partial x}dx\right)dy\right] - \left[\left(a_x(y) + \frac{\partial a_x}{\partial y}\right)dx\right] - [a_y\, dy] \\ &= \left(\frac{\partial a_y}{\partial x} - \frac{\partial a_x}{\partial y}\right)dx\, dy \\ &= (\nabla \times a)\cdot dS \end{aligned}$$

Where, $dS = dxdy\hat{k}.$

Usage

In practice, the above definition is rarely used because in virtually all cases, the curl operator can be applied using some set of curvilinear coordinates, for which simpler representations have been derived.

The notation $\nabla \times F$ has its origins in the similarities to the 3-dimensional cross product, and it is useful as a mnemonic in Cartesian coordinates if ∇ is taken as a vector differential operator del. Such notation involving operators is common in physics and algebra. However, in certain coordinate systems, such as polar-toroidal coordinates (common in plasma physics), using the notation $\nabla \times F$ will yield an incorrect result.

Expanded in Cartesian coordinates, $\nabla \times F$ is, for F composed of $[F_x, F_y, F_z]$:

$$\begin{vmatrix} \mathrm{i} & \mathbf{j} & \mathbf{k} \\ \dfrac{\partial}{\partial x} & \dfrac{\partial}{\partial y} & \dfrac{\partial}{\partial z} \\ F_x & F_y & F_z \end{vmatrix}$$

where i, j, and k are the unit vectors for the *x*-, *y*-, and *z*-axes, respectively. This expands as follows:

$$\left(\frac{\partial F_z}{\partial y} - \frac{\partial F_y}{\partial z}\right)\mathbf{i} + \left(\frac{\partial F_x}{\partial z} - \frac{\partial F_z}{\partial x}\right)\mathbf{j} + \left(\frac{\partial F_y}{\partial x} - \frac{\partial F_x}{\partial y}\right)\mathbf{k}$$

Although expressed in terms of coordinates, the result is invariant under proper rotations of the coordinate axes but the result inverts under reflection.

In a general coordinate system, the curl is given by:

$$(\nabla \times \mathbf{F})^k = \varepsilon^{k\ell m}\nabla_\ell F_m$$

where ε denotes the Levi-Civita tensor and ∇ the covariant derivative, the metric tensor is used to lower the index on F, and the Einstein summation convention implies that repeated indices are summed over. Equivalently,

$$(\nabla \times \mathbf{F}) = \mathbf{e}_k \varepsilon^{k\ell m} \nabla_l F_m$$

where $\mathbf{e}_k$ are the coordinate vector fields. Equivalently, using the exterior derivative, the curl can be expressed as:

$$\nabla \times \mathbf{F} = \left(\star(\mathbf{d}F^\flat)\right)^\sharp$$

Here ♭ and ♯ are the musical isomorphisms, and ★ is the Hodge dual. This formula shows how to calculate the curl of F in any coordinate system, and how to extend the curl to any oriented three-dimensional Riemannian manifold. Since this depends on a choice of orientation, curl is a chiral operation. In other words, if the orientation is reversed, then the direction of the curl is also reversed.

Example

Take the vector field:

$$\mathbf{F}(x, y, z) = y\hat{\mathbf{x}} - x\hat{\mathbf{y}}.$$

Its corresponding plot:

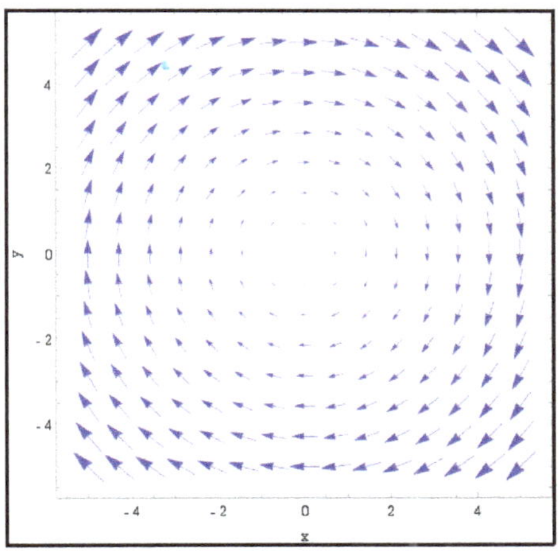

Upon visual inspection, the field can be described as "rotating". If a stationary object were to be placed in the field with the vectors representing a linear force, the object would rotate clockwise.

Calculating the curl:

$$\nabla \times \mathbf{F} = 0\hat{\mathbf{x}} + 0\hat{\mathbf{y}} + \left(\frac{\partial}{\partial x}(-x) - \frac{\partial}{\partial y} y\right)\hat{\mathbf{z}} = -2\hat{\mathbf{z}}$$

The resulting vector field describing the curl would be uniformly going in the negative z direction. It should be noted that the results of this equation align with what could have been predicted using the right-hand rule using a right-handed coordinate system. Being a uniform vector field, the object described before would have the same rotational intensity regardless of where it was placed.

The plot describing the curl of F:

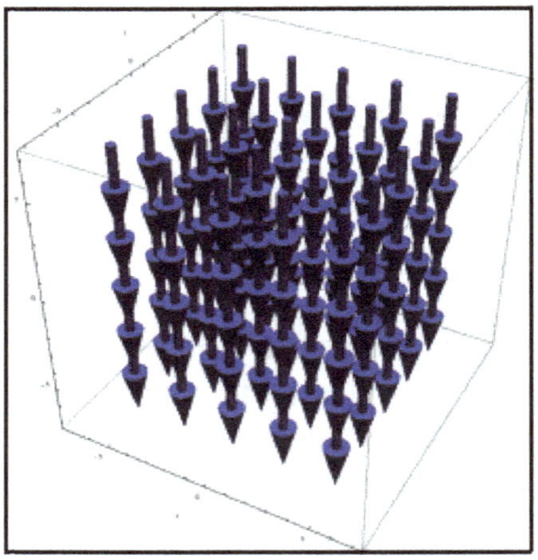

Take the vector field:

$$\mathbf{F}(x, y, z) = -x^2\hat{\mathbf{y}}.$$

Its corresponding plot:

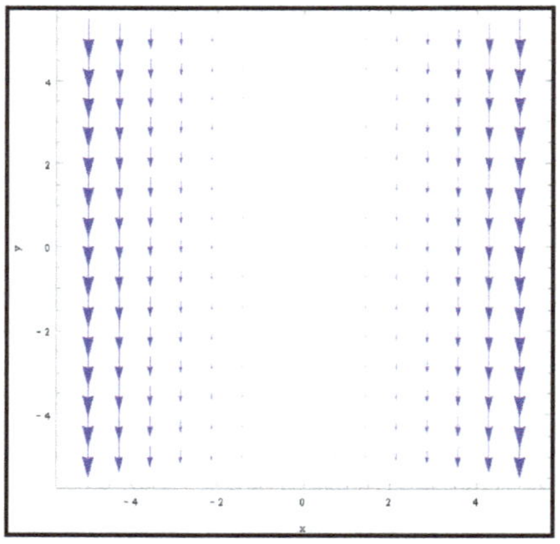

Upon initial inspection, curl existing in this graph would not be obvious. However, taking the object in the previous example, and placing it anywhere on the line $x = 3$, the force exerted on the right side would be slightly greater than the force exerted on the left, causing it to rotate clockwise. Using the right-hand rule, it can be predicted that the resulting curl would be straight in the negative z direction. Inversely, if placed on $x = -3$, the object would rotate counterclockwise and the right-hand rule would result in a positive z direction.

Calculating the curl:

$$\nabla \times \mathbf{F} = 0\hat{\mathbf{x}} + 0\hat{\mathbf{y}} + \frac{\partial}{\partial x}(-x^2)\hat{\mathbf{z}} = -2x\hat{\mathbf{z}}.$$

As predicted, the curl points in the negative z direction when x is positive and vice versa. In this field, the intensity of rotation would be greater as the object moves away from the plane $x = 0$.

The plot describing the curl of **F**:

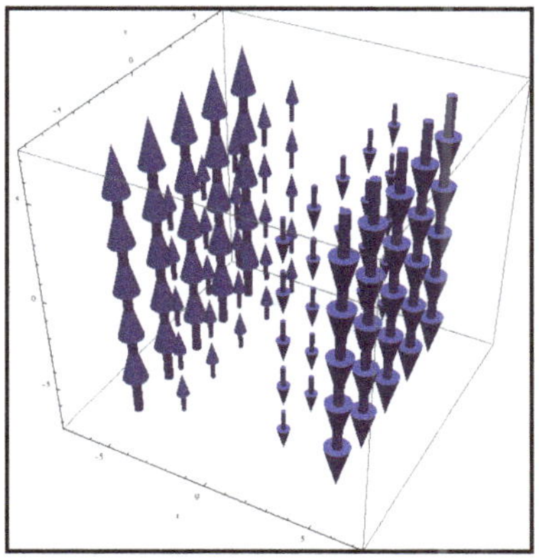

Identities

In general curvilinear coordinates (not only in Cartesian coordinates), the curl of a cross product of vector fields **v** and **F** can be shown to be:

$$\nabla \times (\mathbf{v} \times \mathbf{F}) = \big((\nabla \cdot \mathbf{F}) + \mathbf{F} \cdot \nabla\big)\mathbf{v} - \big((\nabla \cdot \mathbf{v}) + \mathbf{v} \cdot \nabla\big)\mathbf{F}$$

Interchanging the vector field **v** and ∇ operator, we arrive at the cross product of a vector field with curl of a vector field:

$$\mathbf{v} \times (\nabla \times \mathbf{F}) = \nabla_{\mathbf{F}}(\mathbf{v} \cdot \mathbf{F}) - (\mathbf{v} \cdot \nabla)\mathbf{F},$$

where $\nabla_{\mathbf{F}}$ is the Feynman subscript notation, which considers only the variation due to the vector field **F** (i.e., in this case, **v** is treated as being constant in space).

Another example is the curl of a curl of a vector field. It can be shown that in general coordinates

$$\nabla \times (\nabla \times \mathbf{F}) = \nabla(\nabla \cdot \mathbf{F}) - \nabla^2 \mathbf{F},$$

and this identity defines the vector Laplacian of **F**, symbolized as $\nabla^2\mathbf{F}$.

The curl of the gradient of *any* scalar field φ is always the zero vector field:

$$\nabla \times (\nabla \varphi) = \mathbf{0}$$

which follows from the antisymmetry in the definition of the curl, and the symmetry of second derivatives.

If φ is a scalar valued function and **F** is a vector field, then:

$$\nabla \times (\varphi \mathbf{F}) = \nabla \varphi \times \mathbf{F} + \varphi \nabla \times \mathbf{F}$$

Descriptive Examples

- In a vector field describing the linear velocities of each part of a rotating disk, the curl has the same value at all points.

- Of the four Maxwell's equations, two—Faraday's law and Ampère's law—can be compactly expressed using curl. Faraday's law states that the curl of an electric field is equal to the opposite of the time rate of change of the magnetic field, while Ampère's law relates the curl of the magnetic field to the current and rate of change of the electric field.

Generalizations

The vector calculus operations of grad, curl, and div are most easily generalized and understood in the context of differential forms, which involves a number of steps. In a nutshell, they correspond to the derivatives of 0-forms, 1-forms, and 2-forms, respectively. The geometric interpretation of curl as rotation corresponds to identifying bivectors (2-vectors) in 3 dimensions with the special orthogonal Lie algebra $\mathfrak{so}(3)$ of infinitesimal rotations (in coordinates, skew-symmetric 3×3 matrices), while representing rotations by vectors corresponds to identifying 1-vectors (equivalently, 2-vectors) and $\mathfrak{so}(3)$, these all being 3-dimensional spaces.

Differential Forms

In 3 dimensions, a differential 0-form is simply a function $f(x, y, z)$; a differential 1-form is the following expression:

$$a_1\,dx + a_2\,dy + a_3\,dz;$$

a differential 2-form is the formal sum:

$$a_{12}\,dx \wedge dy + a_{13}\,dx \wedge dz + a_{23}\,dy \wedge dz;$$

and a differential 3-form is defined by a single term:

$$a_{123}\,dx \wedge dy \wedge dz.$$

(Here the a-coefficients are real functions; the "wedge products", e.g. $dx \wedge dy$, can be interpreted as some kind of oriented area elements, $dx \wedge dy = -dy \wedge dx$, etc.)

The exterior derivative of a k-form in $\mathbb{R}^3$ is defined as the $(k + 1)$-form from above—and in $\mathbb{R}^n$ if, e.g.,

$$\omega^{(k)} = \sum_{\substack{i_1 < i_2 < \cdots < i_k \\ \forall i_\nu \in 1,\ldots,n}} a_{i_1,\ldots,i_k}\, dx_{i_1} \wedge \cdots \wedge dx_{i_k},$$

then the exterior derivative d leads to,

$$d\omega^{(k)} = \sum_{\substack{j=1 \\ i_1 < \cdots < i_k}}^{n} \frac{\partial a_{i_1,\ldots,i_k}}{\partial x_j} dx_j \wedge dx_{i_1} \wedge \cdots \wedge dx_{i_k}.$$

The exterior derivative of a 1-form is therefore a 2-form, and that of a 2-form is a 3-form. On the other hand, because of the interchangeability of mixed derivatives, e.g. because of,

$$\frac{\partial^2}{\partial x \partial y} = \frac{\partial^2}{\partial y \partial x},$$

the twofold application of the exterior derivative leads to 0.

Thus, denoting the space of k-forms by $\Omega^k(\mathbb{R}^3)$ and the exterior derivative by d one gets a sequence:

$$0 \xrightarrow{d} \Omega^0(\mathbf{R}^3) \xrightarrow{d} \Omega^1(\mathbf{R}^3) \xrightarrow{d} \Omega^2(\mathbf{R}^3) \xrightarrow{d} \Omega^3(\mathbf{R}^3) \xrightarrow{d} 0.$$

Here $\Omega^k(\mathbb{R}^n)$ is the space of sections of the exterior algebra $\Lambda^k(\mathbb{R}^n)$ vector bundle over $\mathbb{R}^n$, whose dimension is the binomial coefficient $\binom{n}{k}$; note that $\Omega^k(\mathbb{R}^3) = 0$ for $k > 3$ or $k < 0$. Writing only dimensions, one obtains a row of Pascal's triangle:

$$0 \to 1 \to 3 \to 3 \to 1 \to 0;$$

the 1-dimensional fibers correspond to functions, and the 3-dimensional fibers to vector fields, as described below. Modulo suitable identifications, the three nontrivial occurrences of the exterior derivative correspond to grad, curl, and div.

Differential forms and the differential can be defined on any Euclidean space, or indeed any manifold, without any notion of a Riemannian metric. On a Riemannian manifold, or more generally pseudo-Riemannian manifold, k-forms can be identified with k-vector fields (k-forms are k-covector fields, and a pseudo-Riemannian metric gives an isomorphism between vectors and covectors), and on an *oriented* vector space with a nondegenerate form (an isomorphism between vectors and covectors), there is an isomorphism between k-vectors and $(n - k)$-vectors; in particular on (the tangent space of) an oriented pseudo-Riemannian manifold. Thus on an oriented pseudo-Riemannian manifold, one can interchange k-forms, k-vector fields, $(n - k)$-forms, and $(n - k)$-vector fields; this is known as Hodge duality. Concretely, on $\mathbb{R}^3$ this is given by:

- 1-forms and 1-vector fields: the 1-form $a_x\, dx + a_y\, dy + a_z\, dz$ corresponds to the vector field (a_x, a_y, a_z).
- 1-forms and 2-forms: one replaces dx by the dual quantity $dy \wedge dz$ (i.e., omit dx), and likewise, taking care of orientation: dy corresponds to $dz \wedge dx = -dx \wedge dz$, and dz corresponds to $dx \wedge dy$. Thus the form $a_x\, dx + a_y\, dy + a_z\, dz$ corresponds to the "dual form" $a_z\, dx \wedge dy + a_y\, dz \wedge dx + a_x\, dy \wedge dz$.

Thus, identifying 0-forms and 3-forms with functions, and 1-forms and 2-forms with vector fields:

- grad takes a function (0-form) to a vector field (1-form);
- curl takes a vector field (1-form) to a vector field (2-form);
- div takes a vector field (2-form) to a function (3-form);

On the other hand, the fact that $d^2 = 0$ corresponds to the identities:

$$\nabla \times (\nabla f) = 0$$

for any function f, and

$$\nabla \cdot (\nabla \times \mathbf{v}) = 0$$

for any vector field **v**.

Grad and div generalize to all oriented pseudo-Riemannian manifolds, with the same geometric interpretation, because the spaces of 0-forms and n-forms is always (fiberwise) 1-dimensional and can be identified with scalar functions, while the spaces of 1-forms and $(n - 1)$-forms are always fiberwise n-dimensional and can be identified with vector fields.

Curl does not generalize in this way to 4 or more dimensions (or down to 2 or fewer dimensions); in 4 dimensions the dimensions are

$$0 \to 1 \to 4 \to 6 \to 4 \to 1 \to 0;$$

so the curl of a 1-vector field (fiberwise 4-dimensional) is a *2-vector field*, which is fiberwise 6-dimensional, one has

$$\omega^{(2)} = \sum_{i<k=1,2,3,4} a_{i,k}\, dx_i \wedge dx_k,$$

which yields a sum of six independent terms, and cannot be identified with a 1-vector field. Nor can one meaningfully go from a 1-vector field to a 2-vector field to a 3-vector field $(4 \to 6 \to 4)$, as taking the differential twice yields zero ($d^2 = 0$). Thus there is no curl function from vector fields to vector fields in other dimensions arising in this way.

However, one can define a curl of a vector field as a *2-vector field* in general, as described below.

Curl Geometrically

2-vectors correspond to the exterior power $\Lambda^2 V$; in the presence of an inner product, in coordinates these are the skew-ymmetric matrices, which are geometrically considered as the special orthogonal Lie algebra $\mathfrak{so}(V)$ of infinitesimal rotations. This has $\binom{n}{2} = \frac{1}{2}n(n-1)$ dimensions, and allows one to interpret the differential of a 1-vector fieldas its infinitesimal rotations. Only in 3 dimensions (or trivially in 0 dimensions) does $n = \frac{1}{2}n(n-1)$, which is the most elegant and common case. In 2 dimensions the curl of a vector field is not a vector field but a function, as 2-dimensional rotations are given by an angle (a scalar – an orientation is required to choose whether one counts clockwise or counterclockwise rotations as positive); this is not the div, but is rather perpendicular to it. In 3 dimensions the curl of a vector field is a vector field as is familiar (in 1 and 0 dimensions the curl of a vector field is 0, because there are no non-trivial 2-vectors), while in 4 dimensions the curl of a vector field is, geometrically, at each point an element of the 6-dimensional Lie algebra $\mathfrak{so}$.

Note also that the curl of a 3-dimensional vector field which only depends on 2 coordinates (say x and y) is simply a vertical vector field (in the z direction) whose magnitude is the curl of the 2-dimensional vector field.

Considering curl as a 2-vector field (an antisymmetric 2-tensor) has been used to generalize vector calculus and associated physics to higher dimensions.

Taking the Velocity Field, v(x,y) , as a Vivid Example for the Geometrical Explanation of Curl

In order to explain curl v in a manner as simple as possible we imagine the vectors of a "whirl-field", which may exist in a fluid, to be parallel to (or to lie directly in) the x,y – plane: v = v(x,y) = (v_x,v_y). And we will stick to this assumption in all our ad-hoc explanations below. What we want to investigate geometrically is the relation,

$v = k \cdot r^n$ (1- a) (vector v being perpendicular to the radius vector r),

which is representing a smoothly increasing (n > 0) or decreasing (n < 0) function "v over r", and which is suitable for discussing those (few) cases which indeed can be met in (engineering) practice. k is a constant of arbitrary value, i.e. we can also write k = 1, which however does not mean that k is left without dimension (!): k [$1/(\sec m^{n-1})$]. Now, "*rot*" is quite generally referred to the rotary motion of an infinitesimally small fluid-element or liquid particle (in short L.P.). In general this motion, i.e. that of an individual particle of the said kind, has 2 components, which remain unchanged here over time:

(a) The 'revolution' around the centre of the (macroscopically extended) whirl, which we locate in the origin of our coordinate system, x = y = 0 = "point zero", so-to-say. This our system, " Σ(x,y) " by designation, is supposed to be at rest relative to the ground; i.e. the circular 'basin' included, within which the (incompressible) fluid moves on circles always parallel to the earth's surface.

(b) The rotation of an envisaged small L.P. (at a certain distance, r, from the centre of the whirl) around its body axis: A 'microscopic' rotation which occurs additionally to its 'macroscopic' revolution around "point zero". While the latter motion is governed by the angular velocity ω_{mac} (r) = v(r) /r, the former one is characterised by ω_{mic} (r). As we will see later on, it depends on n in eq. (1- a) whether ω_{mic} does exist as a separately distinguishable rotational component or not. What should be stressed at last, is that we are establishing here an idealised, model-like experimental arrangement or – better to say – "thought experiment":

Firstly it is assumed that the fluid in the circular basin or trough has a very low viscosity, thus leading to a rather constant circular motion of the fluid over a longer time of observation.

Secondly we replace an unobservable, tiny liquid particle by a visible (though equally small) solid object in the form of a dark coated Paddle Wheel (in short: P.W.), thus enabling us to see any kind of 'elementary' circular motion, be it a macro- or microscopic one. In order to ensure that the P.W. always moves exactly on a circle around the zero point (x=y=0) its axis is pivoted in an arm (of an effective length **r**), which in turn is pivoted at the zero point. So that arm may also be moved manually in addition to its motion, which is normally effected only by the floating P.W.

Thirdly we want to define a *standard situation or case* as a 'point of reference' for certain other cases concerning the P.W.: In the standard case the P.W. is freely floating with the fluid, i.e. without any surplus motion along its circle around the zero point in Σ(x,y). Hence (in the standard or "normal" case) there is no separate or additional (manual) action on the connecting arm (or rod) itself.

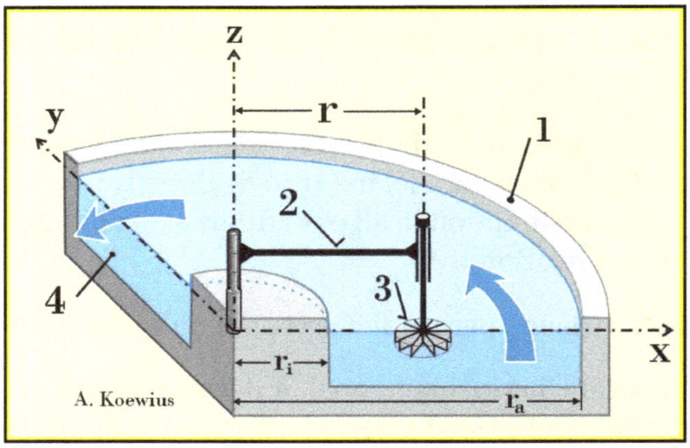

Experimental device suited for the determination of "curl" in a fluid, which is subject to a rotary movement in a circular trough.

1 = section through the circular trough which is fixed on the ground.

2 = connecting arm/rod, pivoted in the centre of the whole device and carrying the Paddle Wheel (3) at its free end. Both 2 and 3 are assumed to be pivoted without any friction.

3 = Paddle Wheel (P.W.), pivoted at the end of the arm (2). The arm is carried along by the P.W. The 'infinitesimally small' P.W. had to be drawn here by far larger with good reason.

4 = a fluid which is circulating in the trough (1) under certain boundary conditions and assumptions concerning v(r), i.e. the velocity distribution over r in the trough between r_a and r_i.

Some Definitions Involving Div, Curl and Grad

- A vector field with zero divergence is said to be solenoidal.
- A vector field with zero curl is said to be irrotational.
- A scalar field with zero gradient is said to be, er, constant.

Gradient Theorem

The gradient theorem, also known as the fundamental theorem of line integrals, is a theorem which states that a line integral taken over a vector field which is the gradient of a scalar function can be evaluated only by looking at the endpoints of the scalar function. In mathematical terms,

$$\int_{p1}^{p2} \nabla f \cdot \vec{dr} = f(p_2) - f(p_1)$$

As a corollary, the line integral over ∇f is path independent, therefore any closed path over ∇f will be equal to zero. Gradient vector fields are also known as conservative.

Proof

Let f be a differentiable function and ∇f be its gradient.

$$\int_{\gamma} \nabla f(u)du = \int_{a}^{b} \nabla f(r(t) \cdot r'(t)dt$$

Since the derivative of f with respect to t will be,

$$\frac{d}{dt}(f(r(t)) = \nabla f(r(t)) \cdot r'(t)$$

by the multivariable chain rule, this expression becomes,

$$\int_{b}^{a} \frac{d}{dt}(f(r(t))dt = f(r(b)) - f(r(a))$$

by the fundamental theorem of calculus.

Example

Suppose $\gamma \subset \mathbb{R}^2$ is the circular arc oriented counterclockwise from (5, 0) to (−4, 3). Using the definition of a line integral,

$$\begin{aligned}
\int_{\gamma} y\mathrm{d}x + x\mathrm{d}y &= \int_{0}^{\pi - \tan^{-1}\left(\frac{3}{4}\right)} ((5\sin t)(-5\sin t) + (5\cos t)(5\cos t))\mathrm{d}t \\
&= \int_{0}^{\pi - \tan^{-1}\left(\frac{3}{4}\right)} 25\left(-\sin^2 t + \cos^2 t\right)\mathrm{d}t \\
&= \int_{0}^{\pi - \tan^{-1}\left(\frac{3}{4}\right)} 25\cos(2t)\,\mathrm{d}t \\
&= \frac{25}{2}\sin(2t)\Big|_{0}^{\pi - \tan^{-1}\left(\frac{3}{4}\right)} \\
&= \frac{25}{2}\sin\left(2\pi - 2\tan^{-1}\left(\frac{3}{4}\right)\right) \\
&= -\frac{25}{2}\sin\left(2\tan^{-1}\left(\frac{3}{4}\right)\right) \\
&= -\frac{25\left(\frac{3}{4}\right)}{\left(\frac{3}{4}\right)^2 + 1} = -12.
\end{aligned}$$

Notice all of the painstaking computations involved in directly calculating the integral. Instead, since the function $f(x, y) = xy$ is differentiable on all of $\mathbb{R}^2$, we can simply use the gradient theorem to say,

$$\int_\gamma y\mathrm{d}x + x\mathrm{d}y = \int_\gamma \nabla(xy)\cdot(\mathrm{d}x, \mathrm{d}y) = xy\,|_{(5,0)}^{(-4,3)} = -4\cdot 3 - 5\cdot 0 = -12.$$

Notice that either way gives the same answer, but using the latter method, most of the work is already done in the proof of the gradient theorem.

Example

For a more abstract example, suppose $\gamma \subset \mathbb{R}^n$ has endpoints p, q, with orientation from p to q. For u in $\mathbb{R}^n$, let |u| denote the Euclidean norm of u. If $\alpha \geq 1$ is a real number, then,

$$\int_\gamma |\mathbf{x}|^{\alpha-1}\,\mathbf{x}\cdot\mathrm{d}\mathbf{x} = \frac{1}{\alpha+1}\int_\gamma (\alpha+1)\,|\mathbf{x}|^{(\alpha+1)-2}\,\mathbf{x}\cdot\mathrm{d}\mathbf{x}$$

$$= \frac{1}{\alpha+1}\int_\gamma \nabla|\mathbf{x}|^{\alpha+1}\cdot\mathrm{d}\mathbf{x} = \frac{|\mathbf{q}|^{\alpha+1} - |\mathbf{p}|^{\alpha+1}}{\alpha+1}$$

Here the final equality follows by the gradient theorem, since the function $f(\mathbf{x}) = |\mathbf{x}|^{\alpha+1}$ is differentiable on $\mathbb{R}^n$ if $\alpha \geq 1$.

If $\alpha < 1$ then this equality will still hold in most cases, but caution must be taken if γ passes through or encloses the origin, because the integrand vector field $|x|^{\alpha-1}x$ will fail to be defined there. However, the case $\alpha = -1$ is somewhat different; in this case, the integrand becomes $|x|^{-2}x = \nabla(\log\ |\mathbf{x}|)$, so that the final equality becomes log |q| – log |p|.

Note that if $n = 1$, then this example is simply a slight variant of the familiar power rule from single-variable calculus.

Suppose there are n point charges arranged in three-dimensional space, and the i-th point charge has charge Q_i and is located at position p_i in $\mathbb{R}^3$. We would like to calculate the work done on a particle of charge q as it travels from a point a to a point b in $\mathbb{R}^3$. Using Coulomb's law, we can easily determine that the force on the particle at position r will be:

$$\mathbf{F}(\mathbf{r}) = kq\sum_{i=1}^{n}\frac{Q_i(\mathbf{r}-\mathbf{p}_i)}{|\mathbf{r}-\mathbf{p}_i|^3}$$

Here |u| denotes the Euclidean norm of the vector u in $\mathbb{R}^3$, and $k = 1/(4\pi\varepsilon_0)$, where ε_0 is the vacuum permittivity.

Let $\gamma \subset \mathbb{R}^3 - \{p_1, \ldots, p_n\}$ be an arbitrary differentiable curve from a to b. Then the work done on the particle is:

$$W = \int_\gamma \mathbf{F}(\mathbf{r})\cdot\mathrm{d}\mathbf{r} = \int_\gamma \left(kq\sum_{i=1}^{n}\frac{Q_i(\mathbf{r}-\mathbf{p}_i)}{|\mathbf{r}-\mathbf{p}_i|^3}\right)\cdot\mathrm{d}\mathbf{r} = kq\sum_{i=1}^{n}\left(Q_i\int_\gamma \frac{\mathbf{r}-\mathbf{p}_i}{|\mathbf{r}-\mathbf{p}_i|^3}\cdot\mathrm{d}\mathbf{r}\right)$$

Now for each i, direct computation shows that,

$$\frac{\mathbf{r}-\mathbf{p}_i}{|\mathbf{r}-\mathbf{p}_i|^3} = -\nabla \frac{1}{|\mathbf{r}-\mathbf{p}_i|}.$$

Thus, continuing from above and using the gradient theorem,

$$W = -kq\sum_{i=1}^{n}\left(Q_i\int_\gamma \nabla\frac{1}{|\mathbf{r}-\mathbf{p}_i|}\cdot d\mathbf{r}\right) = kq\sum_{i=1}^{n}Q_i\left(\frac{1}{|\mathbf{a}-\mathbf{p}_i|}-\frac{1}{|\mathbf{b}-\mathbf{p}_i|}\right)$$

We are finished. Of course, we could have easily completed this calculation using the powerful language of electrostatic potential or electrostatic potential energy (with the familiar formulas $W = -\Delta U = -q\Delta V$). However, we have not yet *defined* potential or potential energy, because the *converse* of the gradient theorem is required to prove that these are well-defined, differentiable functions and that these formulas hold. Thus, we have solved this problem using only Coulomb's Law, the definition of work, and the gradient theorem.

Converse of the Gradient Theorem

The gradient theorem states that if the vector field F is the gradient of some scalar-valued function (i.e., if F is conservative), then F is a path-independent vector field (i.e., the integral of F over some piece-wise-differentiable curve is dependent only on end points). This theorem has a powerful converse:

If F is a path-independent vector field, then F is the gradient of some scalar-valued function.

It is straightforward to show that a vector field is path-independent if and only if the integral of the vector field over every closed loop in its domain is zero. Thus the converse can alternatively be stated as follows: If the integral of F over every closed loop in the domain of F is zero, then F is the gradient of some scalar-valued function.

Proof of the Converse

Suppose U is an open, path-connected subset of $\mathbb{R}^n$, and $F : U \to \mathbb{R}^n$ is a continuous and path-independent vector field. Fix some element a of U, and define $f : U \to \mathbb{R}$ by:

$$f(\mathbf{x}) := \int_{\gamma[\mathbf{a},\mathbf{x}]} \mathbf{F}(\mathbf{u})\cdot d\mathbf{u}$$

Here $\gamma[\mathbf{a}, \mathbf{x}]$ is any (differentiable) curve in U originating at $\mathbf{a}$ and terminating at $\mathbf{x}$. We know that f is well-defined because $\mathbf{F}$ is path-independent.

Let $\mathbf{v}$ be any nonzero vector in $\mathbb{R}^n$. By the definition of the directional derivative,

$$\begin{aligned}\frac{\partial f}{\partial \mathbf{v}}(\mathbf{x}) &= \lim_{t\to 0}\frac{f(\mathbf{x}+t\mathbf{v})-f(\mathbf{x})}{t}\\ &= \lim_{t\to 0}\frac{\int_{\gamma[\mathbf{a},\mathbf{x}+t\mathbf{v}]}\mathbf{F}(\mathbf{u})\cdot d\mathbf{u} - \int_{\gamma[\mathbf{a},\mathbf{x}]}\mathbf{F}(\mathbf{u})\cdot du}{t}\end{aligned}$$

$$= \lim_{t\to 0}\frac{1}{t}\int_{\gamma[\mathbf{x},\mathbf{x}+t\mathbf{v}]}\mathbf{F}(\mathbf{u})\cdot d\mathbf{u}$$

To calculate the integral within the final limit, we must parametrize $\gamma[x, x + tv]$. Since F is path-independent, U is open, and t is approaching zero, we may assume that this path is a straight line, and parametrize it as $\mathbf{u}(s) = \mathbf{x} + s\mathbf{v}$ for $0 < s < t$. Now, since $\mathbf{u}'(s) = \mathbf{v}$, the limit becomes:

$$\lim_{t\to 0}\frac{1}{t}\int_0^t \mathbf{F}(\mathbf{u}(s))\cdot\mathbf{u}'(s)ds = \frac{d}{dt}\int_0^t \mathbf{F}(\mathbf{x}+s\mathbf{v})\cdot\mathbf{v}ds\,\Big|_{t=0} = \mathbf{F}(\mathbf{x})\cdot\mathbf{v}$$

Thus we have a formula for $\partial_{\mathbf{v}} f$, where $\mathbf{v}$ is arbitrary. Let $\mathbf{x} = (x_1, x_2, ..., x_n)$ and let $\mathbf{e}_i$ denote the i-th standard basis vector, so that,

$$\nabla f(\mathbf{x}) = \left(\frac{\partial f(\mathbf{x})}{\partial x_1}, \frac{\partial f(\mathbf{x})}{\partial x_2}, \ldots, \frac{\partial f(\mathbf{x})}{\partial x_n}\right) = (\mathbf{F}(\mathbf{x})\cdot\mathbf{e}_1, \mathbf{F}(\mathbf{x})\cdot\mathbf{e}_2, \ldots, \mathbf{F}(\mathbf{x})\cdot\mathbf{e}_n) = \mathbf{F}(\mathbf{x})$$

Thus we have found a scalar-valued function f whose gradient is the path-independent vector field F, as desired.

Example of the Converse Principle

To illustrate the power of this converse principle, we cite an example that has significant physical consequences. In classical electromagnetism, the electric force is a path-independent force; i.e. the work done on a particle that has returned to its original position within an electric field is zero (assuming that no changing magnetic fields are present).

Therefore, the above theorem implies that the electric force field $F_e : S \to \mathbb{R}^3$ is conservative (here S is some open, path-connected subset of $\mathbb{R}^3$ that contains a charge distribution). Following the ideas of the above proof, we can set some reference point a in S, and define a function $U_e : S \to \mathbb{R}$ by:

$$U_e(\mathbf{r}) := -\int_{\gamma[\mathbf{a},\mathbf{r}]}\mathbf{F}_e(\mathbf{u})\cdot d\mathbf{u}$$

Using the above proof, we know U_e is well-defined and differentiable, and $\mathbf{F}_e = -\nabla U_e$ (from this formula we can use the gradient theorem to easily derive the well-known formula for calculating work done by conservative forces: $W = -\Delta U$). This function U_e is often referred to as the electrostatic potential energy of the system of charges in S (with reference to the zero-of-potential **a**). In many cases, the domain S is assumed to be unbounded and the reference point **a** is taken to be "infinity", which can be made rigorous using limiting techniques. This function U_e is an indispensable tool used in the analysis of many physical systems.

Generalizations

Many of the critical theorems of vector calculus generalize elegantly to statements about the integration of differential forms on manifolds. In the language of differential forms and exterior derivatives, the gradient theorem states that

$$\int_{\partial\gamma}\phi = \int_{\gamma} d\phi$$

for any 0-form, ϕ, defined on some differentiable curve $\gamma \subset \mathbb{R}^n$ (here the integral of ϕ over the boundary of the γ is understood to be the evaluation of at the endpoints of γ).

Notice the striking similarity between this statement and the generalized version of Stokes' theorem, which says that the integral of any compactly supported differential form ω over the boundary of some orientable manifold Ω is equal to the integral of its exterior derivative $d\omega$ over the whole of Ω, i.e.,

$$\int_{\partial\Omega} \omega = \int_{\Omega} d\omega$$

This powerful statement is a generalization of the gradient theorem from 1-forms defined on one-dimensional manifolds to differential forms defined on manifolds of arbitrary dimension.

The converse statement of the gradient theorem also has a powerful generalization in terms of differential forms on manifolds. In particular, suppose ω is a form defined on a contractible domain, and the integral of ω over any closed manifold is zero. Then there exists a form ψ such that $\omega = d\psi$. Thus, on a contractible domain, every closed form is exact. This result is summarized by the Poincaré lemma.

Divergence Theorem

The divergence theorem (also called Gauss's theorem or Gauss-Ostrogradsky theorem) is a theorem which relates the flux of a vector field through a closed surface to the vector field inside the surface. The theorem states that the outward flux of a vector field through a closed surface is equal to the triple integral of the divergence of the vector field inside the surface.

What this theorem actually states is the physical fact that in the absence of the creation or destruction of matter, the density within a region of space can change only by having it flow into or away from the region through its boundary.

The theorem is very applicable in different areas of physics, among others electrostatics and fluid dynamics. Another very important application of the theorem is that several physical laws can be written in both differential and integral form.

Statement: "The volume integral of the divergence of a vector field A taken over any volume Vbounded by a closed surfaceS is equal to the surface integral of A over the surfaceS."

Or

"The volume integral of the divergence of a vector field over the volume enclosed by surface S isequal to the flux of that vector field taken over that surface S."

Hence, this theorem is used to convert volume integral into surface integral.

Under suitable conditions, if E is a region of three dimensional space and D is its boundary surface, oriented outward, then.

$$\iint_D \mathbf{F}\cdot\mathbf{N}dS = \iiint_E \nabla\cdot\mathbf{F}dV.$$

Proof: Again this theorem is too difficult to prove here, but a special case is easier. In the proof of a special case of Green's Theorem, we needed to know that we could describe the region of integration in both possible orders, so that we could set up one double integral using *dxdy* and another using *dydx* . Similarly here, we need to be able to describe the three-dimensional region *E* in different ways.

We start by rewriting the triple integral:

$$\iiint_E \nabla\cdot\mathbf{F}dV = \iiint_E (P_x + Q_y + R_z)dV = \iiint_E P_x dV + \iiint_E Q_y dV + \iiint_E R_z dV.$$

The double integral may be rewritten:

$$\iint_D F\cdot \mathrm{N}dS = \iint_D (Pi + Qj + Rk)\cdot \mathrm{N}dS = \iint_D Pi\cdot \mathrm{N}dS + \iint_D Qj\cdot \mathrm{N}dS + \iint_D Rk\cdot \mathrm{N}dS.$$

To prove that these give the same value it is sufficient to prove that,

$$\iint_D Pi\cdot \mathrm{N}\,dS = \iiint_E P_x dV$$

$$\iint_D Qj\cdot \mathrm{N}\,dS = \iiint_E Q_y dV \quad ,and$$

$$\iint_D Rk\cdot \mathrm{N}\,dS = \iiint_E R_z dV.$$

We set the triple integral up with *dx* dx innermost:

$$\iiint_E P_x dV = \iint_B \int_{g_1(y,z)}^{g_2(y,z)} P_x dxdA = \iint_B P(g_2(y,z),y,z) - P(g_1(y,z),y,z)dA,$$

where *B* is the region in the *y* -*z* plane over which we integrate. The boundary surface of *E* consists of a "top" $x = g_2(y,z)$, a "bottom" $x = g_1(y,z)$, and a "wrap-around side" that is vertical to the *y* -*z* plane. To integrate over the entire boundary surface, we can integrate over each of these (top, bottom, side) and add the results. Over the side surface, the vector N is perpendicular to the vector i, so,

$$\iint_{\text{side}} P\mathbf{i}\cdot\mathbf{N}dS = \iint_{\text{side}} 0dS = 0.$$

Thus, we are left with just the surface integral over the top plus the surface integral over the bottom.

For the top, we use the vector function $\mathbf{r} = \langle g_2(y,z), y, z \rangle$ which gives $\mathbf{r}_y \times \mathbf{r}_z = \langle 1, -g_{2y}, -g_{2z} \rangle$; the dot product of this with $\mathbf{i} = \langle 1,0,0 \rangle$ is 1. Then,

$$\iint_{\text{top}} P\mathbf{i} \cdot \mathbf{N} dS = \iint_B P(g_2(y,z), y, z) dA.$$

In almost identical fashion we get,

$$\iint_{\text{bottom}} P\mathbf{i} \cdot \mathbf{N} dS = -\iint_B P(g_1(y,z), y, z) dA,$$

where the negative sign is needed to make **N** point in the negative x direction. Now,

$$\iint_D P\mathbf{i} \cdot \mathbf{N} dS = \iint_B P(g_2(y,z), y, z) dA - \iint_B P(g_1(y,z), y, z) dA,$$

which is the same as the value of the triple integral above.

Example: Let $F = \langle 2x, 3y, z^2 \rangle$, and consider the three-dimensional volume inside the cube with faces parallel to the principal planes and opposite corners at (0,0,0) and (1,1,1). We compute the two integrals of the divergence theorem.

The triple integral is the easier of the two:

$$\int_0^1 \int_0^1 \int_0^1 2 + 3 + 2z \, dx \, dy \, dz = 6.$$

The surface integral must be separated into six parts, one for each face of the cube. One face is $z=0$ or $\mathbf{r}=\langle u,v,0 \rangle$, $0 \le u,v \le 1$. Then $\mathbf{r}_u = \langle 1,0,0 \rangle$, $\mathbf{r}_v = \langle 0,1,0 \rangle$, and $\mathbf{r}_u \times \mathbf{r}_v = \langle 0,0,1 \rangle$. We need this to be oriented downward (out of the cube), so we use $\langle 0,0,-1 \rangle$ and the corresponding integral is:

$$\int_0^1 \int_0^1 -z^2 \, du \, dv = \int_0^1 \int_0^1 0 \, du \, dv = 0.$$

Another face is $y=1$ or $\mathbf{r}=\langle u,1,v \rangle$. Then $\mathbf{r}_u = \langle 1,0,0 \rangle$, $\mathbf{r}_v = \langle 0,0,1 \rangle$, and $\mathbf{r}_u \times \mathbf{r}_v = \langle 0,-1,0 \rangle$. We need a normal in the positive y direction, so we convert this to $\langle 0,1,0 \rangle$, and the corresponding integral is:

$$\int_0^1 \int_0^1 3y \, du \, dv = \int_0^1 \int_0^1 3 \, du \, dv = 3.$$

The remaining four integrals have values 0, 0, 2, and 1, and the sum of these is 6, in agreement with the triple integral.

Example: Let $F=\langle x^3, y^3, z^2 \rangle$, and consider the cylindrical volume $x^2 + y^2 \le 9$, $0 \le z \le 2$. The triple integral (using cylindrical coordinates) is:

$$\int_0^{2\pi} \int_0^3 \int_0^3 (3r^2 + 2z) r \, dz \, dr \, d\theta = 279\pi.$$

For the surface we need three integrals. The top of the cylinder can be represented by $\mathbf{r}=\langle v\cos u, v\sin u, 2 \rangle$; $\mathbf{r}_u \times \mathbf{r}_v = \langle 0,0,-v \rangle$, which points down into the cylinder, so we convert it to $\langle 0,0,v \rangle$.

Then,

$$\int_0^{2\pi}\int_0^3 \left\langle v^3\cos^3 u, v^3\sin^3 u, 4\right\rangle \cdot \langle 0,0,-v\rangle dvdu = \int_0^{2\pi}\int_0^3 4u\ dvdu = 36\pi.$$

The bottom is r=⟨vcos u,vsin u,0⟩ ; $\mathbf{r}_u \times \mathbf{r}_v$ =⟨0,0,−v⟩ and,

$$\int_0^{2\pi}\int_0^3 \left\langle v^3\cos^3 u, v^3\sin^3 u, 0\right\rangle \cdot \langle 0,0,-v\rangle dvdu = \int_0^{2\pi}\int_0^3 0dvdu = 0.$$

The side of the cylinder is r=⟨3cos u, 3sin u, v⟩; $\mathbf{r}_u \times \mathbf{r}_v$ =⟨3cos u, 3sin u, 0⟩ which does point outward, so,

$$\int_0^{2\pi}\int_0^2 \left\langle 27cos^3 u, 27sin^3 u, v^2\right\rangle \cdot \langle 3cos u, 3sin u, 0\rangle\, dv\, du$$

$$\int_0^{2\pi}\int_0^2 81cos^4 u + 81sin^4 u\, dv\, du = 243\pi.$$

The total surface integral is thus $36\pi + 0 + 243\pi = 279\pi$.

Corollaries

By replacing $\mathbf{F}$ in the divergence theorem with specific forms, other useful identities can be derived (cf. vector identities).

- With $\mathbf{F} \to \mathbf{F}g$ for a scalar function g and a vector field F,

  $$\iiint_V \left[\mathbf{F}\cdot(\nabla g) + g(\nabla\cdot\mathbf{F})\right]dV = \oiint_S g\mathbf{F}\cdot\mathbf{n}dS.$$

  A special case of this is $\mathbf{F} = \nabla f$, in which case the theorem is the basis for Green's identities.

- With $\mathbf{F} \to \mathbf{F}\times\mathbf{G}$ for two vector fields F and G,

  $$\iiint_V \left[\mathbf{G}\cdot(\nabla\times\mathbf{F}) - \mathbf{F}\cdot(\nabla\times\mathbf{G})\right]dV = \oiint_S (\mathbf{F}\times\mathbf{G})\cdot d\mathbf{S}.$$

- With $\mathbf{F} \to f\mathbf{c}$ for a scalar function f and vector field c:

  $$\iiint_V \mathbf{c}\cdot\nabla f\, dV = \oiint_S (\mathbf{c}f)\cdot d\mathbf{S} - \iiint_V f(\nabla\cdot\mathbf{c})dV.$$

  The last term on the right vanishes for constant or any divergence free (solenoidal) vector field, e.g. Incompressible flows without sources or sinks such as phase change or chemical reactions etc.

- With $\mathbf{F} \to \mathbf{c}\times\mathbf{F}$ for vector field F and constant vector c:

  $$\iiint_V \mathbf{c}\cdot(\nabla\times\mathbf{F})dV = \oiint_S (\mathbf{F}\times\mathbf{c})\cdot d\mathbf{S}.$$

Example

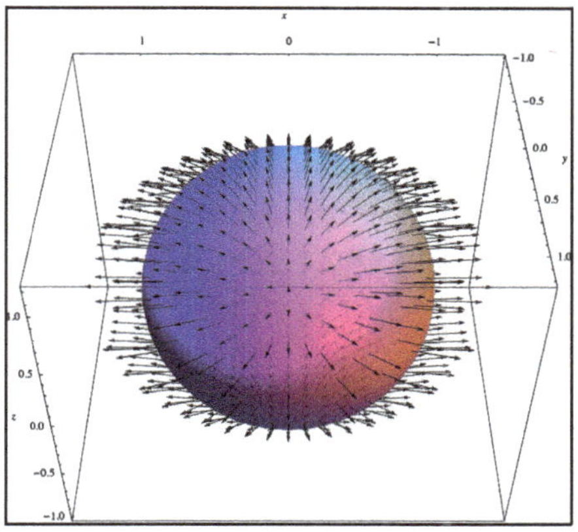

The vector field corresponding to the example shown.

Suppose we wish to evaluate,

$$\oiint_S \mathbf{F} \cdot \mathbf{n} dS,$$

where S is the unit sphere defined by,

$$S = \{(x, y, z) \in \mathbb{R}^3 : x^2 + y^2 + z^2 = 1\},$$

and $\mathbf{F}$ is the vector field,

$$\mathbf{F} = 2x\mathbf{i} + y^2\mathbf{j} + z^2\mathbf{k}.$$

The direct computation of this integral is quite difficult, but we can simplify the derivation of the result using the divergence theorem, because the divergence theorem says that the integral is equal to:

$$\iiint_W (\nabla \cdot \mathbf{F}) dV = 2\iiint_W (1 + y + z) dV = 2\iiint_W dV + 2\iiint_W y dV + 2\iiint_W z dV,$$

where W is the unit ball:

$$W = \{(x, y, z) \in \mathbb{R}^3 : x^2 + y^2 + z^2 \leq 1\}.$$

Since the function y is positive in one hemisphere of W and negative in the other, in an equal and opposite way, its total integral over W is zero. The same is true for z:

$$\iiint_W y dV = \iiint_W z dV = 0.$$

Therefore,

$$\oiint_S \mathbf{F} \cdot \mathbf{n} dS = 2\iiint_W dV = \frac{8\pi}{3},$$

because the unit ball W has volume $\frac{4\pi}{3}$.

Applications

Differential form and Integral form of Physical Laws

As a result of the divergence theorem, a host of physical laws can be written in both a differential form (where one quantity is the divergence of another) and an integral form (where the flux of one quantity through a closed surface is equal to another quantity). Three examples are Gauss's law (in electrostatics), Gauss's law for magnetism, and Gauss's law for gravity.

Continuity Equations

Continuity equations offer more examples of laws with both differential and integral forms, related to each other by the divergence theorem. In fluid dynamics, electromagnetism, quantum mechanics, relativity theory, and a number of other fields, there are continuity equations that describe the conservation of mass, momentum, energy, probability, or other quantities. Generically, these equations state that the divergence of the flow of the conserved quantity is equal to the distribution of *sources* or *sinks* of that quantity. The divergence theorem states that any such continuity equation can be written in a differential form (in terms of a divergence) and an integral form (in terms of a flux).

Inverse-square Laws

Any *inverse-square law* can instead be written in a *Gauss's law*-type form (with a differential and integral form, as described above). Two examples are Gauss's law (in electrostatics), which follows from the inverse-square Coulomb's law, and Gauss's law for gravity, which follows from the inverse-square Newton's law of universal gravitation. The derivation of the Gauss's law-type equation from the inverse-square formulation or vice versa is exactly the same in both cases.

Examples

To verify the planar variant of the divergence theorem for a region R:

$$R = \{(x, y) \in \mathbb{R}^2 : x^2 + y^2 \le 1\},$$

and the vector field:

$$\mathbf{F}(x, y) = 2y\mathbf{i} + 5x\mathbf{j}.$$

The boundary of R is the unit circle, C, that can be represented parametrically by:

$$x = \cos(s), \quad y = \sin(s)$$

such that $0 \le s \le 2\pi$ where s units is the length arc from the point $s = 0$ to the point P on C. Then a vector equation of C is:

$$C(s) = \cos(s)\mathbf{i} + \sin(s)\mathbf{j}.$$

At a point P on C:

$$P = (\cos(s), \sin(s)) \Rightarrow \mathbf{F} = 2\sin(s)\mathbf{i} + 5\cos(s)\mathbf{j}.$$

Therefore,

$$\oint_C \mathbf{F}\cdot\mathbf{n}ds = \int_0^{2\pi} (2\sin(s)\mathbf{i} + 5\cos(s)\mathbf{j})\cdot(\cos(s)\mathbf{i} + \sin(s)\mathbf{j})ds$$
$$= \int_0^{2\pi} (2\sin(s)\cos(s) + 5\sin(s)\cos(s))ds$$
$$= 7\int_0^{2\pi} \sin(s)\cos(s)ds$$
$$= 0.$$

Because $M = 2y$, $\dfrac{\partial M}{\partial x} = 0$, and because $N = 5x, \dfrac{\partial N}{\partial x} = 0$. Thus,

$$\iint_R \nabla\cdot\mathbf{F}dA = \iint_R \left(\frac{\partial M}{\partial x} + \frac{\partial N}{\partial y}\right)dA = 0.$$

Generalizations

Multiple Dimensions

One can use the general Stokes' Theorem to equate the n-dimensional volume integral of the divergence of a vector field $\mathbf{F}$ over a region U to the $(n - 1)$-dimensional surface integral of $\mathbf{F}$ over the boundary of U:

$$\underbrace{\int\cdots\int_U}_{n} \nabla\cdot\ \mathbf{F}\cdot dv = \underbrace{\oint\cdots\oint_{\partial U}}_{n-1} \mathbf{F}\cdot nds$$

This equation is also known as the Divergence theorem.

When $n = 2$, this is equivalent to Green's theorem.

When $n = 1$, it reduces to the Fundamental theorem of calculus.

Tensor Fields

Writing the theorem in Einstein notation:

$$\iiint_V \frac{\partial \mathbf{F}_i}{\partial x_i}dV = \oiint_S \mathbf{F}_i n_i\, dS$$

suggestively, replacing the vector field $\mathbf{F}$ with a rank-n tensor field T, this can be generalized to:

$$\iiint_V \frac{\partial T_{i_1 i_2\cdots i_q\cdots i_n}}{\partial x_{i_q}}dV = \oiint_S T_{i_1 i_2\cdots i_q\cdots i_n} n_{i_q}\, dS.$$

where on each side, tensor contraction occurs for at least one index. This form of the theorem is still in 3d, each index takes values 1, 2, and 3. It can be generalized further still to higher (or lower) dimensions (for example to 4d spacetime in general relativity).

Stokes' Theorem

Stokes' theorem can be regarded as a higher-dimensional version of Green's Theorem. If we recall from previous lessons, Green's Theorem relates a double integral over a plane region to a line integral around its plane boundary curve. Stoke's Theorem relates a surface integral over a surface to a line integral along the boundary curve.

In fact, Stokes' Theorem provides insight into a physical interpretation of the curl. In a vector field, the rotation of the vector field is at a maximum when the curl of the vector field and the normal vector have the same direction. In other words, while the tendency to rotate will vary from point to point on the surface, Stokes' Theorem says that the collective measure of this rotational tendency taken over the entire surface is equal to the tendency of a fluid to circulate around the boundary curve.

Let S be a surface in space and the boundary of S is simple closed curve c. Let $\vec{F}\ (x, y, z)$ is a continuous function that has continuous partial derivatives in S, then,

$$\int_S \left(\vec{\nabla} \times \vec{F}\right) \cdot \hat{n} ds = \oint_C \vec{F} \cdot d\vec{r}$$ where $\hat{n}$ is an outward drawn normal to the elemental surface ds and $d\vec{r}$ is taken along C.

Proof of Stokes' Theorem

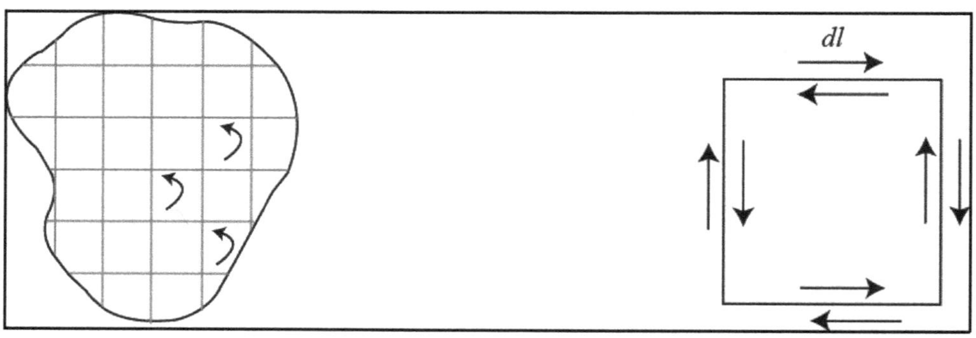

We have shown that circulation around a small mesh is written as,

$$\sum_{4\,sides} \vec{A} . \vec{dl} = \left(\vec{\nabla} \times \vec{A}\right) dxdy$$

The surface integrals (i.e. RHS of the above equation) can be added together. Again (as in the divergence theorem case) the line integrals of the interior line segments cancel identically. Only the integral around the perimeter survives, giving:

$$\sum_{external\ line\ segments} \vec{A}.\vec{dl} = \sum_{rectangle}\left(\vec{\nabla} \times \vec{A}\right).d\vec{s}$$

Then, $\oint \vec{A}.\vec{dl} = \int\left(\vec{\nabla} \times \vec{A}\right).d\vec{s}$

Example

Verify Stokes' theorem for $\vec{F}\,(2x - y)\hat{i} - yz^2\hat{j} - y^2z\hat{k}$ for the paraboloid S devoted by:

$$z = f\,(x, y) = 1 - (x^2 + y^2) \qquad z \geq 0$$

Or the upper half surface of a sphere.

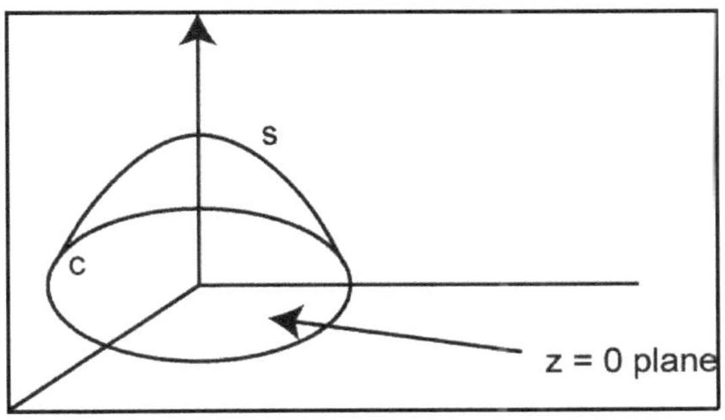

In z = o plane the boundary c of the surface S is a circle $(x^2 + y^2) = 1$.

A convenient way to determine the line integral (refer to Stoke's theorem) is to substitute.

$x = \cos t$, $y = sint$. $0 \leq t \leq 2\pi$ and $z = 0$.

Thus,

$$\oint_c \vec{F}.\vec{dr} = \oint_c (2x - y).(dx\,i + dy\,j + dz\,\hat{k})$$

$$= \oint_c (2x - y)dx = \int_0^{2\pi} 2\cos t - \sin t)(-\sin t)\,dt$$

$$= \pi$$

Also,

$$\vec{\nabla} \times \vec{F} = \hat{K}(\hat{z}), so\left(\vec{\nabla} \times \vec{F}\right).d\vec{s} = \int \hat{z}.\hat{n}\,ds$$

$$\int dxdy = \int_{x=-1}^{1} \int_{y=-\sqrt{1-x^2}}^{\sqrt{1-x^2}} dxdy$$

$= \pi$. (verified)

Examples

Some example based on Stokes' theorem are given below:

Example 1: If F(x, y, z) = xi + zj + 2yk and curve C is obtained by the intersection of circle $x^2 + y^2 = 1$ by the plane z = x, oriented upward. Verify Stokes' theorem.

Solution: Given that $x^2 + y^2 = 1$ and plane z = x

$$r(t) = < cos\ t,\ sin\ t,\ cos\ t >$$

$$dr = < -sin\ t,\ cos\ t,\ -sin\ t > dt$$

F(x, y, z) = xi + zj + 2yk

$$F(t) = < cos\ t,\ cos\ t,\ 2\ sin\ t >$$

F . dr = (-sin t cos t + $\cos^2$t - 2 $\sin^2$ t) dt

$$\int_C F.dr = \int_0^{2\pi} (-sintcost + cos^2t - 2sin^2t)dt$$

$$= \int_0^{2\pi} \left(\frac{-1}{2} \sin 2t + 1 - 3\sin^2 t \right) dt$$

$$= \int_0^{2\pi} \left(\frac{-1}{2} \sin 2t + 1 - \frac{3}{2}(1 - \cos 2t \right) dt$$

$$= \int_0^{2\pi} \left(\frac{-1}{2} \sin\ 2t + \frac{3}{2} \cos\ 2t - \frac{1}{2} \right) dt$$

$$= \left[\frac{1}{4} \cos\ 2t + \frac{3}{4} \sin 2t - \frac{1}{2} t \right]$$

$$= \left(\frac{1}{4} \cos\ 4\pi + \frac{3}{4} \sin\ 4\pi - \pi \right) - \left(\frac{1}{4} \right)$$

$$= -\pi$$

$$F = < x,\ z,\ 2y >$$

$$curl\ F = < 1,\ 0,\ 0 >$$

$$dS = < -1,\ 0,\ 1 > dx\ dy$$

$$\iint_C curl\, F.dS = \iint_{unit\ disc} <1,0,0>.<-1,0,1>\, dx\, dy$$

$$= -\iint_{unit\ disc} dx\, dy$$

$$= -\pi$$

It verifies that $\int_C F.dr = \int_C curl\, F.dS$ which means Stokes' theorem holds.

Example2: Calculate the line integral given by $\int_C F.r(s)ds$, if $F = yi + xz^3 j - zy^3 k$ and curve C is given by the circle $x^2 + y^2 = 4$ and $z = -3$.

Solution: In Stokes' theorem, the normal vector n points towards the direction of positive z; therefore n = k.

We are going to evaluate curl $F = \nabla \times F$

$$\begin{vmatrix} i & j & k \\ \frac{\partial}{\partial x} & \frac{\partial}{\partial y} & \frac{\partial}{\partial z} \\ y & xz^3 & -zy^3 \end{vmatrix}$$

$$= -3z(y^2 + xz)i + (z^3 - 1)k$$

$$= <-3z(y^2 + xz), 0, (z^3 - 1)>$$

$$curl\, F.\, n = <-3z(y^2 + xz), 0, (z^3 - 1)>.<0,0,1> \ (\text{since n=k})$$

$$= z^3 - 1$$

$$[curl\, F.\, n]_{z=-3} = -27 - 1 = -28$$

$$\iint_{x^2+y^2=4, z=-3} curl\, F.\, n = -28\, dx\, dy$$

$$= -28(2^2)\pi$$

$$= -112\pi$$

Formulation for Smooth Manifolds with Boundary

Let Ω be an oriented smooth manifold with boundary of dimension n and let α be a smooth n-differential form that is compactly supported on Ω. First, suppose that α is compactly supported in the domain of a single, oriented coordinate chart $\{U, \varphi\}$. In this case, we define the integral of α over Ω as:

$$\int_\Omega \alpha = \int_{\varphi(U)} \left(\varphi^{-1}\right)^* \alpha,$$

i.e., via the pullback of α to R^n.

More generally, the integral of α over Ω is defined as follows: Let $\{\psi_i\}$ be a partition of unity associated with a locally finitecover $\{U_i, \varphi_i\}$ of (consistently oriented) coordinate charts, then define the integral:

$$\int_{\Omega} \alpha \equiv \sum_i \int_{U_i} \psi_i \alpha,$$

where each term in the sum is evaluated by pulling back to R^n as described above. This quantity is well-defined; that is, it does not depend on the choice of the coordinate charts, nor the partition of unity.

The generalized Stokes' theorem reads:

Theorem: *(Stokes–Cartan) If ω is a smooth $(n-1)$-form with compact support on smooth n-dimensional manifold-with-boundary Ω, $\partial\Omega$, denotes the boundary of Ω given the induced orientation, and $i:\partial\Omega \hookrightarrow \Omega$ is the inclusion map, then,*

$$\int_{\Omega} d\omega = \int_{\partial\Omega} i^* \omega.$$

Conventionally, $\int_{\partial\Omega} i^*\omega$ is abbreviated as $\int_{\partial\Omega} \omega$,, since the pullback of a differential form by the inclusion map is simply its restriction to its domain: $i^*\omega = \omega|_{\partial\Omega}$. Here d is the exterior derivative, which is defined using the manifold structure only. The right-hand side is sometimes written as $\oint_{\partial\Omega} \omega$ to stress the fact that the $(n-1)-$ manifold $\partial\Omega$ has no boundary. (This fact is also an implication of Stokes' theorem, since for a given smooth n-dimensional manifold Ω,, application of the theorem twice gives $\int_{\partial(\partial\Omega)} \omega = \int_{\Omega} d(d\omega) = 0$ for any $(n-2)$-form ω, which implies that $\partial(\partial\Omega) = \varnothing$.) The right-hand side of the equation is often used to formulate *integral* laws; the left-hand side then leads to equivalent *differential* formulations.

The theorem is often used in situations where Ω, is an embedded oriented submanifold of some bigger manifold, often $\mathbf{R}^k$, on which the form ω is defined.

Topological Preliminaries; Integration over Chains

Let *M* be a smooth manifold. A (smooth) singular *k*-simplex in *M* is defined as a smooth map from the standard simplex in R^k to *M*. The group $C_k(M, Z)$ of singular *k*-chains on *M* is defined to be the free abelian group on the set of singular *k*-simplices in *M*. These groups, together with the boundary map, ∂, define a chain complex. The corresponding homology (resp. cohomology) group is isomorphic to the usual singular homology group $H_k(M, Z)$ (resp. the singular cohomology group $H^k(M, \mathbf{Z})$), defined using continuous rather than smooth simplices in *M*.

On the other hand, the differential forms, with exterior derivative, *d*, as the connecting map, form a cochain complex, which defines the de Rham cohomology groups $H^k_{dR}(M, R)$.

Differential *k*-forms can be integrated over a *k*-simplex in a natural way, by pulling back to R^k. Extending by linearity allows one to integrate over chains. This gives a linear map from the space of *k*-forms to the *k*th group of singular cochains, $C^k(M, Z)$, the linear functionals on $C^k(M, Z)$. In other words, a *k*-form ω defines a functional.

$$I(\omega)(c) = \oint_c \omega$$

on the k-chains. Stokes' theorem says that this is a chain map from de Rham cohomology to singular cohomology with real coefficients; the exterior derivative, d, behaves like the *dual* of ∂ on forms. This gives a homomorphism from de Rham cohomology to singular cohomology. On the level of forms, this means:

1. closed forms, i.e., $d\omega = 0$, have zero integral over *boundaries*, i.e. over manifolds that can be written as $\partial \Sigma_c M_c$, and
2. exact forms, i.e., $\omega = d\sigma$, have zero integral over *cycles*, i.e. if the boundaries sum up to the empty set: $\Sigma_c M_c = \varnothing$.

De Rham's theorem shows that this homomorphism is in fact an isomorphism. So the converse to 1 and 2 above hold true. In other words, if $\{c_i\}$ are cycles generating the kth homology group, then for any corresponding real numbers, $\{a_i\}$, there exist a closed form, ω, such that,

$$\oint_{c_i} \omega = a_i,$$

and this form is unique up to exact forms.

Stokes' theorem on smooth manifolds can be derived from Stokes' theorem for chains in smooth manifolds, and vice versa.Formally stated, the latter reads:

Theorem: *(Stokes' theorem for chains) If c is a smooth k-chain in a smooth manifold M, and ω is a smooth $(k-1)$– form on M, then,*

$$\int_{\partial c} \omega = \int_c d\omega.$$

Underlying Principle

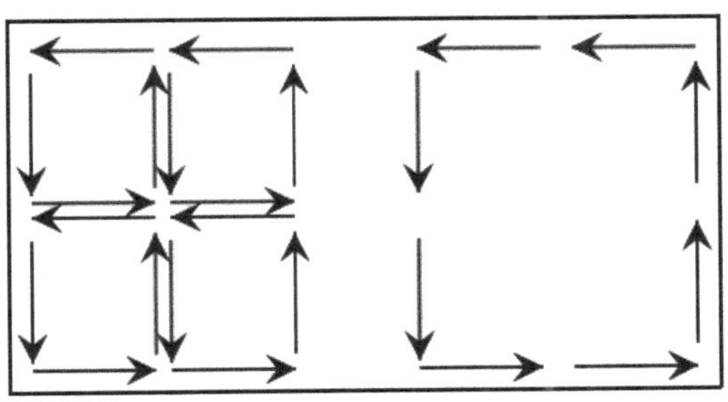

To simplify these topological arguments, it is worthwhile to examine the underlying principle by considering an example for $d = 2$ dimensions. The essential idea can be understood by the diagram on the left, which shows that, in an oriented tiling of a manifold, the interior paths are traversed in opposite directions; their contributions to the path integral thus cancel each other pairwise. As a consequence, only the contribution from the boundary remains. It thus suffices to prove Stokes' theorem for sufficiently fine tilings (or, equivalently, simplices), which usually is not difficult.

Generalization to Rough Sets

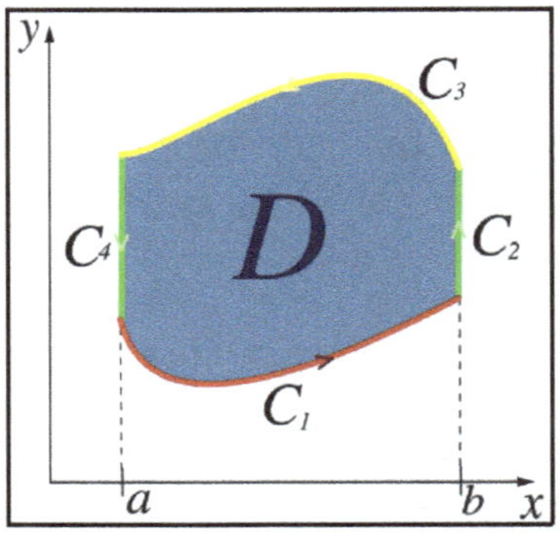

A region (here called Dinstead of Ω) with piecewise smooth boundary. This is a manifold with corners, so its boundary is not a smooth manifold.

The formulation above, in which Ω is a smooth manifold with boundary, does not suffice in many applications. For example, if the domain of integration is defined as the plane region between two x-coordinates and the graphs of two functions, it will often happen that the domain has corners. In such a case, the corner points mean that Ω is not a smooth manifold with boundary, and so the statement of Stokes' theorem given above does not apply. Nevertheless, it is possible to check that the conclusion of Stokes' theorem is still true. This is because Ω and its boundary are well-behaved away from a small set of points (a measure zero set).

A version of Stokes' theorem that extends to rough domains was proved by Whitney.Assume that D is a connected bounded open subset of R^n. Call D a *standard domain* if it satisfies the following property: There exists a subset P of ∂D, open in ∂D, whose complement in ∂D has Hausdorff $(n - 1)$-measure zero; and such that every point of P has a *generalized normal vector*. This is a vector $\mathbf{v}(x)$ such that, if a coordinate system is chosen so that $\mathbf{v}(x)$ is the first basis vector, then, in an open neighborhood around x, there exists a smooth function $f(x_2, ..., x_n)$ such that P is the graph $\{ x_1 = f(x_2, ..., x_n) \}$ and D is the region $\{ x_1 < f(x_2, ..., x_n) \}$. Whitney remarks that the boundary of a standard domain is the union of a set of zero Hausdorff $(n - 1)$-measure and a finite or countable union of smooth $(n - 1)$-manifolds, each of which has the domain on only one side. He then proves that if D is a standard domain in R^n, ω is an $(n - 1)$-form which is defined, continuous, and bounded on $D \cup P$, smooth on D, integrable on P, and such that $d\omega$ is integrable on D, then Stokes' theorem holds, that is,

$$\int_P \omega = \int_D d\omega.$$

The study of measure-theoretic properties of rough sets leads to geometric measure theory. Even more general versions of Stokes' theorem have been proved by Federer and by Harrison.

Special Cases

The general form of the Stokes' theorem using differential forms is more powerful and easier to use than the special cases. The traditional versions can be formulated using Cartesian coordinates without the machinery of differential geometry, and thus are more accessible. Further, they are older and their names are more familiar as a result. The traditional forms are often considered more convenient by practicing scientists and engineers but the non-naturalness of the traditional formulation becomes apparent when using other coordinate systems, even familiar ones like spherical or cylindrical coordinates. There is potential for confusion in the way names are applied, and the use of dual formulations.

Kelvin–Stokes' Theorem

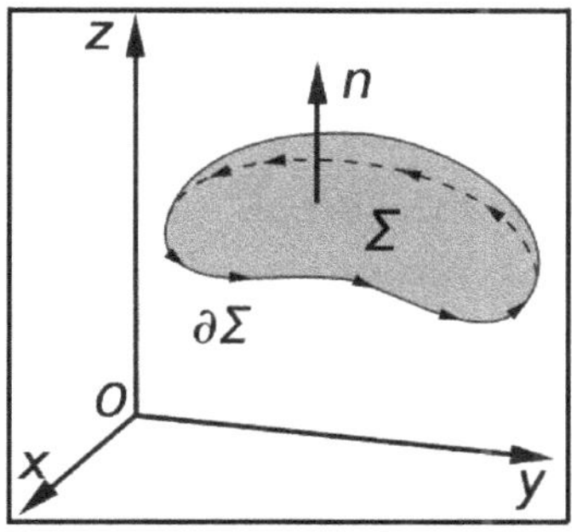

An illustration of the Kelvin–Stokes' theorem, with surface Σ, its boundary ∂Σ and the "normal" vector *n*.

This is a (dualized) (1 + 1)-dimensional case, for a 1-form (dualized because it is a statement about vector fields). This special case is often just referred to as *Stokes' theorem* in many introductory university vector calculus courses and as used in physics and engineering. It is also sometimes known as the curl theorem.

The classical Kelvin–Stokes' theorem relates the surface integral of the curl of a vector field over a surface Σ in Euclidean three-space to the line integral of the vector field over its boundary, is a special case of the general Stokes' theorem (with $n = 2$) once we identify a vector field with a 1-form using the metric on Euclidean 3-space. The curve of the line integral, ∂Σ, must have positive orientation, meaning that ∂Σ points counterclockwise when the surface normal, n, points toward the viewer.

One consequence of the Kelvin–Stokes' theorem is that the field lines of a vector field with zero curl cannot be closed contours. The formula can be rewritten as:

Suppose $\mathbf{F}(P(x,y,z),Q(x,y,z),R(x,y,z))$ is defined in region with smooth surface Σ and has first-order continuous partial derivatives. Then,

$$\iint_{\Sigma}\left(\left(\frac{\partial R}{\partial y}-\frac{\partial Q}{\partial z}\right)dy\,dz+\left(\frac{\partial P}{\partial z}-\frac{\partial R}{\partial x}\right)dz\,dx+\left(\frac{\partial Q}{\partial x}-\frac{\partial P}{\partial y}\right)dx\,dy\right)$$
$$=\oint_{\partial\Sigma}\left(P\,dx+Q\,dy+R\,dz\right),$$

where *P*, *Q* and *R* are the components of F, and $\partial\Sigma$ is boundary of region with smooth surface Σ.

Green's Theorem

Green's theorem gives a relationship between the line integral of a two-dimensional vector field over a closed path in the plane and the double integral over the region it encloses. The fact that the integral of a (two-dimensional) conservative field over a closed path is zero is a special case of Green's theorem.

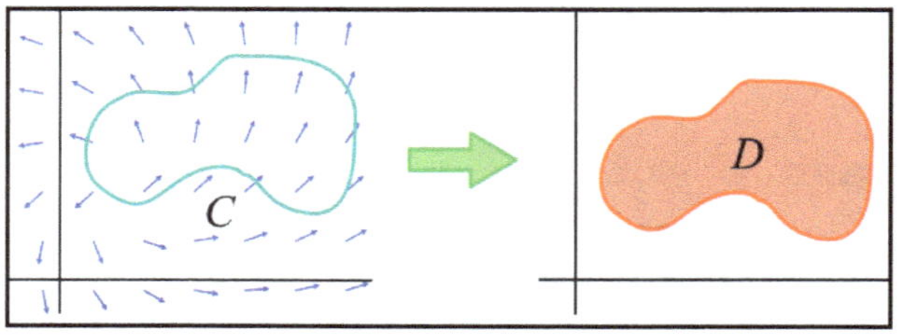

Green's theorem is itself a special case of the much more general Stokes' theorem. The statement in Green's theorem that two different types of integrals are equal can be used to compute either type: sometimes Green's theorem is used to transform a line integral into a double integral, and sometimes it is used to transform a double integral into a line integral.

Proof when D is a Simple Region

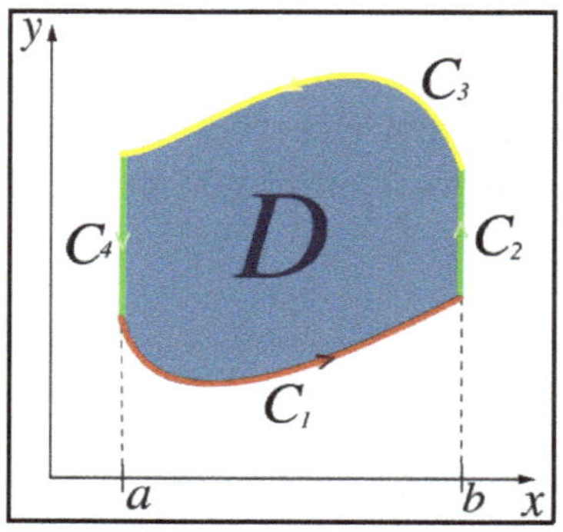

If D is a simple region with its boundary consisting of the curves C_1, C_2, C_3, C_4, half of Green's theorem can be demonstrated.

The following is a proof of half of the theorem for the simplified area D, a type I region where C_1 and C_3 are curves connected by vertical lines (possibly of zero length). A similar proof exists for the other half of the theorem when D is a type II region where C_2 and C_4 are curves connected by horizontal lines (again, possibly of zero length). Putting these two parts together, the theorem is thus proven for regions of type III (defined as regions which are both type I and type II). The general case can then be deduced from this special case by decomposing D into a set of type III regions.

If it can be shown that,

$$\oint_C L\,dx = \iint_D \left(-\frac{\partial L}{\partial y}\right) dA$$

and,

$$\oint_C M\,dy = \iint_D \left(\frac{\partial M}{\partial x}\right) dA$$

are true, then Green's theorem follows immediately for the region D. We can prove easily for regions of type I, and regions of type II. Green's theorem then follows for regions of type III.

Assume region D is a type I region and can thus be characterized, as pictured on the right, by,

$$D = \{(x, y) \mid a \le x \le b, g_1(x) \le y \le g_2(x)\}$$

where g_1 and g_2 are continuous functions on $[a, b]$. Compute the double integral inabove equation:

$$\iint_D \frac{\partial L}{\partial y} dA = \int_a^b \int_{g_1(x)}^{g_2(x)} \frac{\partial L}{\partial y}(x, y) dy dx$$

$$= \int_a^b \{L(x, g_2(x)) - L(x, g_1(x))\} dx.$$

Now compute the line integral in (1). C can be rewritten as the union of four curves: C_1, C_2, C_3, C_4. With C_1, use the parametric equations: $x = x$, $y = g_1(x)$, $a \le x \le b$. Then,

$$\int_{C_1} L(x, y) dx = \int_a^b L(x, g_1(x)) dx.$$

With C_3, use the parametric equations: $x = x$, $y = g_2(x)$, $a \le x \le b$. Then,

$$\int_{C_3} L(x, y) dx = -\int_{-C_3} L(x, y) dx = -\int_a^b L(x, g_2(x)) dx.$$

The integral over C_3 is negated because it goes in the negative direction from b to a, as C is oriented positively (anticlockwise). On C_2 and C_4, x remains constant, meaning:

$$\int_{C_4} L(x, y) dx = \int_{C_2} L(x, y) dx = 0.$$

Therefore,

$$\int_C L dx = \int_{C_1} L(x, y) dx + \int_{C_2} L(x, y) dx + \int_{C_3} L(x, y) dx + \int_{C_4} L(x, y) dx$$

$$= -\int_a^b L(x, g_2(x)) dx + \int_a^b L(x, g_1(x)) dx.$$

Combining above equations, we get regions of type I. A similar treatment yields for regions of type II. Putting the two together, we get the result for regions of type III.

Proof for Rectifiable Jordan Curves

We are going to prove the following

Theorem. Let Γ be a rectifiable, positively oriented Jordan curve in $\mathbf{R}^2$ and let $\mathbf{R}$ denote its inner region. Suppose that $A, B : \overline{R} \to \mathbf{R}$ are continuous functions with the property that A has second partial derivative at every point of R, B has first partial derivative at every point of R and that the functions $D_1 B, D_2 A : R \to \mathbf{R}$ are Riemann-integrable over R.

Then,

$$\int_{\Gamma} Adx + Bdy = \int_{R} (D_1B(x,y) - D_2A(x,y))d(x,y).$$

We need the following lemmas:

Lemma 1 (Decomposition Lemma). Assume Γ is a rectifiable, positively oriented Jordan curve in the plane and let R be its inner region. For every positive real δ, let $\mathcal{F}(\delta)$ denote the collection of squares in the plane bounded by the lines $x = m\delta, y = m\delta,$, where m runs through the set of integers. Then, for this δ, there exists a decomposition of $\bar{R}$ into a finite number of non-overlapping subregions in such a manner that,

(i) Each one of the subregions contained in R, say $R_1, R_2, \ldots, R_k$, is a square from $\mathcal{F}(\delta)$.

(ii) Each one of the remaining subregions, say $R_{k+1}, \ldots, R_s$, has as boundary a rectifiable Jordan curve formed by a finite number of arcs of Γ and parts of the sides of some square from $\mathcal{F}(\delta)$.

(iii) Each one of the border regions R_k $, \ldots, R_s$ can be enclosed in a square of edge-length 2δ.

(iv) If Γ_i is the positively oriented boundary curve of R_i, then $\Gamma = \Gamma_1 + \Gamma_2 + \cdots + \Gamma_s$.

(v) The number $s - k$ of border regions is no greater than $4\left(\frac{\Lambda}{\delta} + 1\right)$, where Λ is the length of Γ.

Lemma 2. Let Γ. be a rectifiable curve in the plane and let $\Delta_{\Gamma}(h)$ be the set of points in the plane whose distance from (the range of) Γ. is at most h. The outer Jordan content of this set satisfies $\bar{c}\,\Delta_{\Gamma}(h) \leq 2h\Lambda + \pi h^2$.

Lemma 3. Let Γ. be a rectifiable curve in $\mathbf{R}^2$ and let f : range of $\Gamma \to \mathbf{R}$ be a continuous function. Then,

$\left|\int_{\Gamma} f(x,y)dy\right|$ and

$\left|\int_{\Gamma} f(x,y)dx\right|$ are $\leq \frac{1}{2}\Lambda\Omega_f$, where Ω_f is the oscillation of f on the range of Γ.

Now we are in position to prove the Theorem:

Proof of Theorem. Let ε be an arbitrary positive real number. By continuity of A, B and compactness of $\bar{R}$, given $\varepsilon > 0$, there exists $0 < \delta < 1$ such that whenever two points of $\bar{R}$ are less than $2\sqrt{2}\delta$ apart, their images under A, B are less than ε apart. For this δ, consider the decomposition given by the previous Lemma. We have,

$$\int_{\Gamma} Adx + Bdy = \sum_{i=1}^{k} \int_{\Gamma_i} Adx + Bdy \quad + \sum_{i=k+1}^{s} \int_{\Gamma_i} Adx + Bdy.$$

Put $\varphi := D_1B - D_2A$.

For each $i \in \{1,\ldots,k\}$, the curve Γ_i is a positively oriented square, for which Green's formula holds. Hence,

$$\sum_{i=1}^{k} \int_{\Gamma_i} Adx + Bdy = \sum_{i=1}^{k} \int_{R_i} \varphi = \int_{\bigcup_{i=1}^{k} R_i} \varphi.$$

Every point of a border region is at a distance no greater than $2\sqrt{2}\delta$ from Γ. Thus, if K is the union of all border regions, then $K \subset \Delta_\Gamma(2\sqrt{2}\delta)$; ; hence $c(K) \leq \bar{c}\Delta_\Gamma(2\sqrt{2}\delta) \leq 4\sqrt{2}\delta + 8\pi\delta^2$, by Lemma 2. Notice that,

$\int_R \varphi - \int_{\bigcup_{i=1}^{k} R_i} \varphi = \int_K \varphi.$ This yields,

$$\left| \sum_{i=1}^{k} \int_{\Gamma_i} Adx + Bdy \quad - \int_R \varphi \right| \leq M\delta(1 + \pi\sqrt{2}\delta) \text{ for some } M > 0.$$

We may as well choose δ so that the RHS of the last inequality is $< \varepsilon$.

The remark in the beginning of this proof implies that the oscillations of A and B on every border region is at most ε. We have,

$$\left| \sum_{i=k+1}^{s} \int_{\Gamma_i} Adx + Bdy \right| \leq \frac{1}{2} \varepsilon \sum_{i=k+1}^{s} \Lambda_i.$$

By Lemma 1(iii),

$$\sum_{i=k+1}^{s} \Lambda_i \leq \Lambda + (4\delta)4\left(\frac{\Lambda}{\delta} + 1\right) \leq 17\Lambda + 16.$$

Combining these, we finally get,

$$\left| \int_\Gamma Adx + Bdy \quad - \int_R \varphi \right| < C\varepsilon,$$

for some $C > 0$. Since this is true for every $\varepsilon > 0$, we are done.

Validity under Different Hypothesis

The hypothesis of the last theorem are not the only ones under which Green's formula is true. Another common set of conditions is the following:

The functions $A, B : \overline{R} \to \mathbf{R}$ are still assumed to be continuous. However, we now require them to be Fréchet-differentiable at every point of R. This implies the existence of all directional derivatives, in particular $D_{e_i} A =: D_i A, D_{e_i} B =: D_i B, i = 1, 2$, where, as usual, (e_1, e_2) is the canonical ordered basis of $\mathbf{R}^2$. In addition, we require the function $D_1 B - D_2 A$ to be Riemann-integrable over R.

It suffices to prove this for squares which are contained in R and have sides parallel to the axes.

The proof then follows the lines of the method employed to prove the Cauchy-Goursat Theorem for triangles.

As a corollary of this, we get the Cauchy Integral Theorem for rectifiable Jordan curves:

Theorem (Cauchy): If Γ is a rectifiable Jordan curve in $\mathbf{C}$ and if f : closure of inner region of $\Gamma \to \mathbf{C}$ is a continuous mapping holomorphic throughout the inner region of Γ, then,

$$\int_\Gamma f = 0,$$

the integral being a complex contour integral.

Proof. We regard the complex plane as $\mathbf{R}^2$. Now, define $u, v : \overline{R} \to \mathbf{R}$ to be such that $f(x+iy) = u(x,y) + iv(x,y)$. These functions are clearly continuous. It is well-known that u and v are Fréchet-differentiable and that they satisfy the Cauchy-Riemann equations: $D_1 v + D_2 u = D_1 u - D_2 v =$ zero function.

Now, analysing the sums used to define the complex contour integral in question, it is easy to realize that,

$$\int_\Gamma f = \int_\Gamma u\,dx - v\,dy \quad + i\int_\Gamma v\,dx + u\,dy,$$

the integrals on the RHS being usual line integrals. These remarks allow us to apply Green's Theorem to each one of these line integrals, finishing the proof.

Measure-theoretic Assumptions

Green's formula also holds when, besides continuity assumptions,

(i) The functions $D_i A, D_i B, i = 1, 2,$, are defined at every point of R, with the exception of a countable subset.

(ii) The function $D_1 B - D_2 A$ is Lebesgue-integrable over R.

Multiply-connected Regions

Theorem. Let $\Gamma_0, \Gamma_1, \ldots, \Gamma_n$ be positively oriented rectifiable Jordan curves in $\mathbf{R}^2$ satisfying:

$$\Gamma_i \subset R_0, \qquad \text{if } 1 \le i \le n$$
$$\Gamma_i \subset \mathbf{R}^2 \setminus \overline{R}_j, \qquad \text{if } 1 \le i, j \le n \text{ and } i \ne j,$$

where R_i is the inner region of Γ_i. Let,

$$D = R_0 \setminus (\overline{R}_1 \cup \overline{R}_2 \cup \cdots \cup \overline{R}_n).$$

Suppose $p : \overline{D} \to \mathbf{R}$ and $q : \overline{D} \to \mathbf{R}$ are continuous functions whose restriction to D is Fréchet-differentiable. If the function.

$$(x,y) \mapsto \frac{\partial q}{\partial e_1}(x,y) - \frac{\partial p}{\partial e_2}(x,y)$$

is Riemann-integrable over D, then,

$$\int_{\Gamma_0} p(x,y)dx + q(x,y)dy - \sum_{i=1}^{n} \int_{\Gamma_i} p(x,y)dx + q(x,y)dy$$
$$= \int_D \left\{ \frac{\partial q}{\partial e_1}(x,y) - \frac{\partial p}{\partial e_2}(x,y) \right\} d(x,y).$$

Relationship to Stokes' Theorem

Green's theorem is a special case of the Kelvin–Stokes' theorem, when applied to a region in the *xy*-plane:

We can augment the two-dimensional field into a three-dimensional field with a *z* component that is always 0. Write F for the vector-valued function $\mathbf{F} = (L, M, 0)$. Start with the left side of Green's theorem:

$$\oint_C (L\,dx + M\,dy) = \oint_C (L, M, 0) \cdot (dx, dy, dz) = \oint_C \mathbf{F} \cdot d\mathbf{r}.$$

The Kelvin–Stokes' theorem:

$$\oint_C \mathbf{F} \cdot d\mathbf{r} = \iint_S \nabla \times \mathbf{F} \cdot \hat{\mathbf{n}}\, dS.$$

The surface S is just the region in the plane D, with the unit normals $\hat{\mathbf{n}}$ pointing up (in the positive *z* direction) to match the "positive orientation" definitions for both theorems.

The expression inside the integral becomes:

$$\nabla \times \mathbf{F} \cdot \hat{\mathbf{n}} = \left[\left(\frac{\partial 0}{\partial y} - \frac{\partial M}{\partial z} \right)\mathbf{i} + \left(\frac{\partial L}{\partial z} - \frac{\partial 0}{\partial x} \right)\mathbf{j} + \left(\frac{\partial M}{\partial x} - \frac{\partial L}{\partial y} \right)\mathbf{k} \right] \cdot \mathbf{k} = \left(\frac{\partial M}{\partial x} - \frac{\partial L}{\partial y} \right).$$

Thus we get the right side of Green's theorem,

$$\iint_S \nabla \times \mathbf{F} \cdot \hat{\mathbf{n}}\, dS = \iint_D \left(\frac{\partial M}{\partial x} - \frac{\partial L}{\partial y} \right) dA.$$

Green's theorem is also a straightforward result of the general Stokes' theorem using differential forms and exterior derivatives:

$$\oint_C L\,dx + M\,dy = \oint_{\partial D} \omega = \int_D d\omega$$
$$= \int_D \frac{\partial L}{\partial y} dy \wedge dx + \frac{\partial M}{\partial x} dx \wedge dy = \iint_D \left(\frac{\partial M}{\partial x} - \frac{\partial L}{\partial y} \right) dx\, dy.$$

Relationship to the Divergence Theorem

Considering only two-dimensional vector fields, Green's theorem is equivalent to the two-dimensional version of the divergence theorem:

$$\iiint_V (\nabla\cdot\mathbf{F})dV = \oiint_{\partial v} (\mathbf{F}\cdot\mathbf{n})dS.$$

where $\nabla\cdot\mathbf{F}$ is the divergence on the two-dimensional vector field $\mathbf{F}$, and $\hat{\mathbf{n}}$ is the outward-pointing unit normal vector on the boundary.

To see this, consider the unit normal $\hat{\mathbf{n}}$ in the right side of the equation. Since in Green›s theorem $d\mathbf{r} = (dx, dy)$ is a vector pointing tangential along the curve, and the curve C is the positively oriented (i.e. anticlockwise) curve along the boundary, an outward normal would be a vector which points 90° to the right of this; one choice would be $(dy,-dx)$. The length of this vector is $\sqrt{dx^2+dy^2} = ds$. So $(dy,-dx) = \hat{\mathbf{n}}ds$.

Start with the left side of Green's theorem:

$$\oint_C (L\,dx + M\,dy) = \oint_C (M,-L)\cdot(dy,-dx) = \oint_C (M,-L)\cdot\hat{\mathbf{n}}ds.$$

Applying the two-dimensional divergence theorem with $\mathbf{F} = (M,-L),$, we get the right side of Green's theorem:

$$\oint_C (M,-L)\cdot\hat{\mathbf{n}}ds = \iint_D (\nabla\cdot(M,-L))\,dA = \iint_D \left(\frac{\partial M}{\partial x} - \frac{\partial L}{\partial y}\right)dA.$$

Area Calculation

Green's theorem can be used to compute area by line integral. The area of D is given by,

$$A = \iint_D dA.$$

Then if we choose L and M such that $\frac{\partial M}{\partial x} - \frac{\partial L}{\partial y} = 1$, the area is given by,

$$A = \oint_C (L\,dx + M\,dy).$$

Possible formulas for the area of D include,

$$A = \oint_C x\,dy = -\oint_C y\,dx = \tfrac{1}{2}\oint_C (-y\,dx + x\,dy).$$

Jacobian Matrix and Determinant

In vector calculus, the Jacobian matrix is the matrix of all first-order partial derivatives of a vector-valued function. Specifically, suppose is a function. Such a function is given by m real-valued component functions, . The partial derivatives of all these functions with respect to the variables can be organized in an m-by-n matrix, the Jacobian matrix of, as follows: This matrix, whose entries are functions of, is also denoted by and . The Jacobian matrix is important because if the function F is differentiable at a point, which is a slightly stronger condition than merely requiring that all partial derivatives exist there, then the derivative of F at p is the linear transformation represented by the matrix . This linear transformation is the best linear approximation of the function F near the point p. In the case the Jacobian matrix will be a square matrix, and its determinant, a function of, is the Jacobian determinant of F. It carries important information about the local behavior of F and can be thought of as a local expansion factor for volumes; it is used when performing variable substitutions in multi-variable integrals, since it occurs prominently in the substitution rule for multiple variables.

Jacobian Matrix

The Jacobian Matrix is just a matrix that takes the partial derivatives of each element of a transformation. In general, the Jacobian Matrix of a transformation F, looks like this:

$$J = \begin{bmatrix} \frac{\partial F_1}{\partial x_1} \cdots \frac{\partial F_1}{\partial x_n} \\ \vdots \quad \ddots \vdots \\ \frac{\partial F_m}{\partial x_1} \cdots \frac{\partial F_m}{\partial x_n} \end{bmatrix} \quad F_1, F_2, F_3 \ldots$$

are each of the elements of the output vector and x_1, x_2, x_3 ... are each of the elements of the input vector.

So for example, in a 2 dimensional case, let T be a transformation such that $T(u,v) = <x, y>$ then the Jacobian matrix of this function would look like this:

$$J(u,v) = \begin{bmatrix} \frac{\partial x}{\partial u} & \frac{\partial x}{\partial v} \\ \frac{\partial y}{\partial u} & \frac{\partial y}{\partial v} \end{bmatrix}$$

This Jacobian matrix noticably holds all of the partial derivatives of the transformation with respect to each of the variables. Therefore each row contains how a particular output element changes with respect to each of the input elements. This means that the Jacobian matrix contains vectors that help describe how a change in any of the input elements affects the output elements.

To help illustrate making Jacobian matrices, let's do some examples:

Example 1:

What would be the Jacobian Matrix of this Transformation?

$$T(u,v) = < u \times cosv, u \times sinv > .$$

Solution:

$$x = u \times cos\, v \rightarrow \frac{\partial x}{\partial u} = cos\, v, \frac{\partial x}{\partial v} = -u \times sin\, v$$

$$y = u \times sin\, v \rightarrow \frac{\partial y}{\partial u} = sin\, v, \frac{\partial y}{\partial v} = u \times cos\, v$$

$$J(u,v) = \begin{bmatrix} \frac{\partial x}{\partial u} & \frac{\partial x}{\partial v} \\ \frac{\partial y}{\partial u} & \frac{\partial y}{\partial v} \end{bmatrix} = \begin{bmatrix} cos\, v & -u \times sin\, v \\ sin\, v & u \times cos\, v \end{bmatrix}$$

Example 2:

What would be the Jacobian Matrix of this Transformation?

$$T(u,v) = < u, v, u^v >, u > 0\ .$$

Solution:

Notice, that this matrix will not be square because there is a difference in dimensions of the input and output, i.e. the transformation is not injective.

$$x = u \rightarrow \frac{\partial x}{\partial u} = 1, \frac{\partial x}{\partial v} = 0$$

$$y = v \rightarrow \frac{\partial x}{\partial u} = 0, \frac{\partial x}{\partial v} = 1$$

$$z = u^v \rightarrow \frac{\partial y}{\partial u} = u^{v-1} \times v, \frac{\partial y}{\partial v} = u^v \times ln(u)$$

$$J(u,v) = \begin{bmatrix} \frac{\partial x}{\partial u} & \frac{\partial x}{\partial v} \\ \frac{\partial y}{\partial u} & \frac{\partial y}{\partial v} \\ \frac{\partial z}{\partial u} & \frac{\partial z}{\partial v} \end{bmatrix} = \begin{bmatrix} 1 & 0 \\ 0 & 1 \\ u^v \times v & u^v \times ln(u) \end{bmatrix}$$

Example 3:

What would be the Jacobian Matrix of this Transformation?

$$T(u,v) = < tan(uv) >\ .$$

Solution:

Notice, that this matrix will not be square because there is a difference in dimensions of the input and output, i.e. the transformation is not injective.

$$x = tan(uv) \rightarrow \frac{\partial x}{\partial u} = sec^2(uv) \times v, \frac{\partial x}{\partial v} = sec^2(uv) \times u$$

$$(u,v) = \begin{bmatrix} \frac{\partial x}{\partial u} & \frac{\partial x}{\partial v} \end{bmatrix} = [sec^2(uv) \times v \; sec^2(uv) \times u]$$

Jacobian Determinant

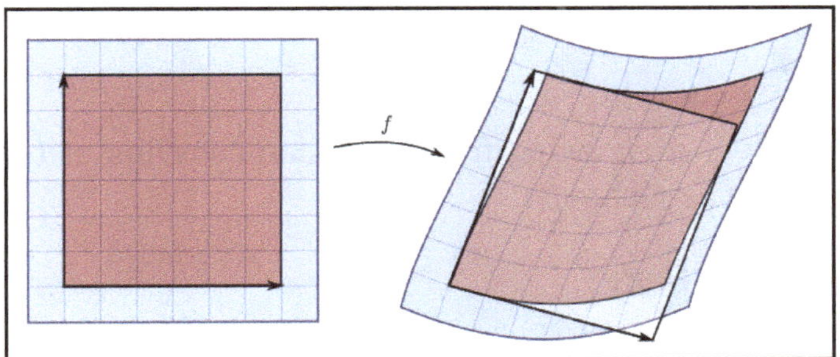

A nonlinear map $f : \mathbb{R}^2 \rightarrow \mathbb{R}^2$ sends a small square to a distorted parallelogram, shown in red. The Jacobian at a point gives the best linear approximation of the parallelogram near that point, shown as a translucent white.

If $m = n$, then **f** is a function from $\mathbb{R}^n$ to itself and the Jacobian matrix is a square matrix. We can then form its determinant, known as the Jacobian determinant. The Jacobian determinant is sometimes referred to as "the Jacobian".

The Jacobian determinant at a given point gives important information about the behavior of f near that point. For instance, the continuously differentiable function f is invertible near a point p ∈ $\mathbb{R}^n$ if the Jacobian determinant at p is non-zero. This is the inverse function theorem. Furthermore, if the Jacobian determinant at p is positive, then f preserves orientation near p; if it is negative, freverses orientation. The absolute value of the Jacobian determinant at p gives us the factor by which the function f expands or shrinks volumes near p; this is why it occurs in the general substitution rule.

The Jacobian determinant is used when making a change of variables when evaluating a multiple integral of a function over a region within its domain. To accommodate for the change of coordinates the magnitude of the Jacobian determinant arises as a multiplicative factor within the integral. This is because the n-dimensional dV element is in general a parallelepiped in the new coordinate system, and the n-volume of a parallelepiped is the determinant of its edge vectors.

The Jacobian can also be used to solve systems of differential equations at an equilibrium point or approximate solutions near an equilibrium point.

Generalizations of the Jacobian Determinant

The Jacobian determinant can be generalized also to the case where the dimension of the target differs from that of the domain. More precisely, let f, U , n, m, and y be as above:

- If $m < n$, the Jacobian of f at y is given by the square root of the determinant of $Df_y \cdot (Df_y)^t$ (where Df_y^t denotes the transpose of the matrix Df_y);

- If $m > n$, the Jacobian of f at y is given by the square root of the determinant of $\left(Df_y\right)^t \cdot Df_y$.

These generalizations play a key role respectively in the Coarea formula and Area formula.

An important characterization of the Jacobian is then given by the Cauchy Binet formula: $Jf\left(y\right)^2$ (is the sum of the squares of the determinants of all $n \times n$ minors of $Df\big|_y$.

$$\begin{bmatrix} cosv & -u & \times sin\,v \\ sinv & u & \times cos\,v \end{bmatrix} = u\cos^2 v + u\sin^2 v = u$$

Notice that, in an integral when changing from cartesian coordinates ($dxdy$) to polar coordinates ($drd\theta$), the equation is as such:

$$dxdy = r \times drd\theta = u \times dudv$$

It is easy to extrapolate, then, that the transformation from one set of coordinates to another set is merely:

$$dC2 = det(J(T))dC1$$

where C1 is the first set of coordinates, det $\left(J\left(C1\right)\right)$ is the determinant of the Jacobian matrix made from the Transformation T, T is the Transformation from C1 to C2 and C2 is the second set of coordinates.

It is important to notice several aspects: first, the determinant is assumed to exist and be non-zero, and therefore the Jacobian matrix must be square and invertible. This makes sense because when changing coordinates, it should be possible to change back.

Moreover, recall from linear algebra that, in a two dimensional case, the determinant of a matrix of two vectors describes the area of the parallelogram drawn by it, or more accurately, it describes the scale factor between the unit square and the area of the parallelogram. If we extend the analogy, the determinant of the Jacobian would describe some sort of scale factor change from one set of coordinates to the other. Here is a picture that should help:

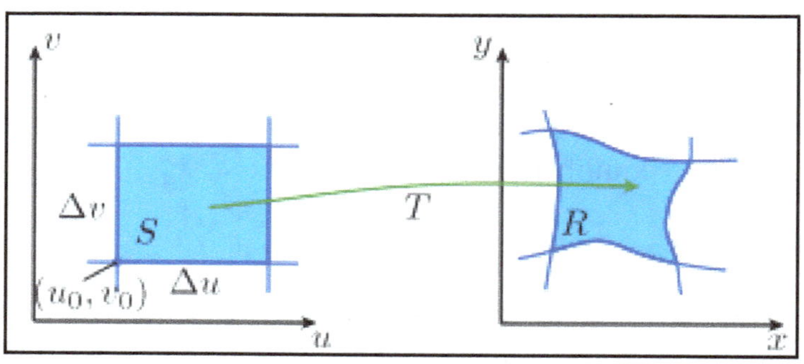

This is the general idea behind change of variables. It is easy to that the jacobian method matches up with one-dimensional change of variables.

$$T(u) = u^2 = x, \quad J(u) = \left[\frac{\partial x}{\partial u}\right] = [2u], \qquad du = |J(u)| \times du = 2u \times dx$$

Remember that when substituting, to be VERY careful. It should always be the case where,

$$J(\vec{u}) \times d\vec{x} = d\vec{u}, \quad J(\vec{x}\ ) \times d\vec{u} = d\vec{x}\ .$$

Inverse

According to the inverse function theorem, the matrix inverse of the Jacobian matrix of an invertible function is the Jacobian matrix of the *inverse* function. That is, if the Jacobian of the function $\mathbf{f}: \mathbb{R}^n \to \mathbb{R}^n$ is continuous and nonsingular at the point $\mathbf{p}$ in $\mathbb{R}^n$, then $\mathbf{f}$ is invertible when restricted to some neighborhood of $\mathbf{p}$ and,

$$\mathbf{J}_{\mathbf{f}^{-1}} \circ \mathbf{f} = \mathbf{J}_{\mathbf{f}}^{-1}.$$

Conversely, if the Jacobian determinant is not zero at a point, then the function is *locally invertible* near this point, that is, there is a neighbourhood of this point in which the function is invertible.

The (unproved) Jacobian conjecture is related to global invertibility in the case of a polynomial function, that is a function defined by n polynomials in n variables. It asserts that, if the Jacobian determinant is a non-zero constant (or, equivalently, that it does not have any complex zero), then the function is invertible and its inverse is a polynomial function.

Critical Points

If $\mathbf{f}: \mathbb{R}^n \to \mathbb{R}^m$ is a differentiable function, a *critical point* of $\mathbf{f}$ is a point where the rank of the Jacobian matrix is not maximal. This means that the rank at the critical point is lower than the rank at some neighbour point. In other words, let k be the maximal dimension of the open balls contained in the image of $\mathbf{f}$; then a point is critical if all minors of rank k of $\mathbf{f}$ are zero.

In the case where $1 = m = n = k$, a point is critical if the Jacobian determinant is zero.

Example 1:

Consider the function $\mathbf{f}: \mathbb{R}^2 \to \mathbb{R}^2$ given by,

$$\mathbf{f}(x, y) = \begin{bmatrix} x^2 y \\ 5x + \sin y \end{bmatrix}.$$

Then we have,

$$f_1(x, y) = x^2 y$$

and,

$$f_2(x, y) = 5x + \sin y$$

and the Jacobian matrix of **F** is:

$$\mathbf{J_f}(x,y)=\begin{bmatrix}\frac{\partial f_1}{\partial x} & \frac{\partial f_1}{\partial y}\\ \frac{\partial f_2}{\partial x} & \frac{\partial f_2}{\partial y}\end{bmatrix}=\begin{bmatrix}2xy & x^2\\ 5 & \cos y\end{bmatrix}$$

and the Jacobian determinant is:

$$\det(\mathbf{J_f}(x,y))=2xy\cos y-5x^2.$$

Example 2: Polar-Cartesian Transformation:

The transformation from polar coordinates (r, φ) to Cartesian coordinates (x, y), is given by the function $\mathbf{F}\colon \mathbb{R}^+ \times [0, 2\pi) \to \mathbb{R}^2$ with components:

$$x = r\cos\varphi;$$
$$y = r\sin\varphi.$$

$$\mathbf{J_F}(r,\varphi)=\begin{bmatrix}\frac{\partial x}{\partial r} & \frac{\partial x}{\partial \varphi}\\ \frac{\partial y}{\partial r} & \frac{\partial y}{\partial \varphi}\end{bmatrix}=\begin{bmatrix}\cos\varphi & -r\sin\varphi\\ \sin\varphi & r\cos\varphi\end{bmatrix}$$

The Jacobian determinant is equal to r. This can be used to transform integrals between the two coordinate systems:

$$\iint_{\mathbf{F}(A)} f(x,y)dx\,dy=\iint_A f(r\cos\varphi, r\sin\varphi)r\,drd\varphi.$$

Example 3: Spherical-Cartesian Transformation:

The transformation from spherical coordinates (r, θ, φ) to Cartesian coordinates (x, y, z), is given by the function $\mathbf{F}\,\mathbb{R}^+ \times [0,\pi] \times [0, 2\pi) \to \mathbb{R}^3$ with components:

$$= r\sin\theta\cos\varphi;$$
$$y = r\sin\theta\sin\varphi;$$
$$z = r\cos\theta.$$

The Jacobian matrix for this coordinate change is:

$$\mathrm{J_F}(r,\theta,\varphi)=\begin{bmatrix}\frac{\partial x}{\partial r} & \frac{\partial x}{\partial \theta} & \frac{\partial x}{\partial \varphi}\\ \frac{\partial y}{\partial r} & \frac{\partial y}{\partial \theta} & \frac{\partial y}{\partial \varphi}\\ \frac{\partial z}{\partial r} & \frac{\partial z}{\partial \theta} & \frac{\partial z}{\partial \varphi}\end{bmatrix}=\begin{bmatrix}\sin\theta\cos\varphi & r\cos\theta\cos\varphi & -r\sin\theta\sin\varphi\\ \sin\theta\sin\varphi & r\cos\theta\sin\varphi & r\sin\theta\cos\varphi\\ \cos\theta & -r\sin\theta & 0\end{bmatrix}.$$

The determinant is $r^2 \sin\theta$. As an example, since $dV = dxdydz$ this determinant implies that the differential volume element $dV = r^2 sin\theta dr d\theta d\varphi$.. Unlike for a change of Cartesian coordinates, this determinant is not a constant, and varies with coordinates (r and θ).

Example 4:

The Jacobian matrix of the function $F : \mathbb{R}^3 \to \mathbb{R}^4$ with components:

$$y_1 = x_1$$
$$y_2 = 5x_3$$
$$y_3 = 4x_2^2 - 2x_3$$
$$y_4 = x_3 \sin x_1$$

is,

$$\mathbf{J}_{\mathbf{F}}(x_1, x_2, x_3) = \begin{bmatrix} \frac{\partial y_1}{\partial x_1} & \frac{\partial y_1}{\partial x_2} & \frac{\partial y_1}{\partial x_3} \\ \frac{\partial y_2}{\partial x_1} & \frac{\partial y_2}{\partial x_2} & \frac{\partial y_2}{\partial x_3} \\ \frac{\partial y_3}{\partial x_1} & \frac{\partial y_3}{\partial x_2} & \frac{\partial y_3}{\partial x_3} \\ \frac{\partial y_4}{\partial x_1} & \frac{\partial y_4}{\partial x_2} & \frac{\partial y_4}{\partial x_3} \end{bmatrix} = \begin{bmatrix} 1 & 0 & 0 \\ 0 & 0 & 5 \\ 0 & 8x_2 & -2 \\ x_3 \cos x_1 & 0 & \sin x_1 \end{bmatrix}.$$

This example shows that the Jacobian need not be a square matrix.

Example 5:

The Jacobian determinant of the function $\mathbf{F} : \mathbb{R}^3 \to \mathbb{R}^3$ with components,

$$y_1 = 5x_2$$
$$y_2 = 4x_1^2 - 2\sin(x_2 x_3)$$
$$y_3 = x_2 x_3$$

is,

$$\begin{vmatrix} 0 & 5 & 0 \\ 8x_1 & -2x_3 \cos(x_2 x_3) & -2x_2 \cos(x_2 x_3) \\ 0 & x_3 & x_2 \end{vmatrix} = -8x_1 \begin{vmatrix} 5 & 0 \\ x_3 & x_2 \end{vmatrix} = -40x_1 x_2.$$

From this we see that F reverses orientation near those points where x_1 and x_2 have the same sign; the function is locally invertible everywhere except near points where $x_1 = 0$ or $x_2 = 0$. Intuitively, if one starts with a tiny object around the point (1, 2, 3) and apply F to that object, one will get a resulting object with approximately $40 \times 1 \times 2 = 80$ times the volume of the original one, with orientation reversed.

Example 6:

Compute the following expression:

$$\iint (x^2 + y^2)dA$$

where dA is the region bounded by $y^2 - x^2 = 9$, $y^2 - x^2 = 16$, $2xy = 14$, $2xy = 6$.

Solution:

First, let's graph the region.

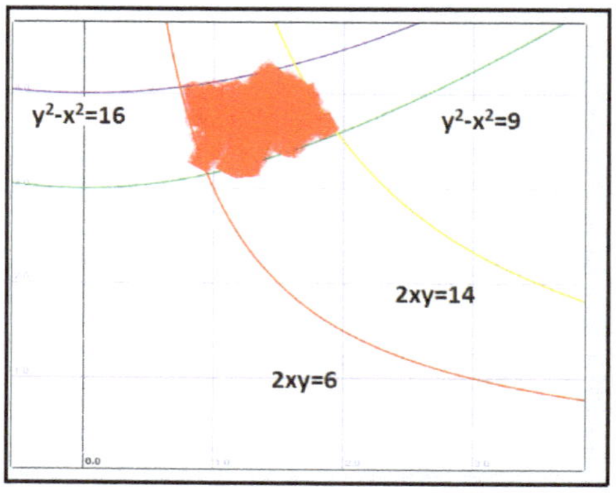

$$Let \quad v = 2xy, \quad u = y^2 - x^2$$

Why do we do this? Let's look at the bounds: they're from $y^2 - x^2 = 9$ to $y^2 - x^2 = 16$ and $2xy = 6$ and $2xy = 14$. It'd be quite a simple task if the integral looked something like this:

$$\int_9^{16}\int_6^{14} k(u,v)dudv$$

However, notice that in this case, instead of substituting x= and y=, we're substituing with u= and v=. Therefore the Jacobian will be written as $J(u,v)$ and the expression will be $dudv = J(u,v)dxdy$.

$$u = 2xy \rightarrow \frac{\partial u}{\partial x} = 2y, \frac{\partial u}{\partial y} = 2x$$

$$y^2 - x^2 \rightarrow \frac{\partial v}{\partial x} = -2x, \frac{\partial v}{\partial y} = 2y$$

$$J(u,v) = \begin{bmatrix} \frac{\partial u}{\partial x} & \frac{\partial u}{\partial y} \\ \frac{\partial v}{\partial x} & \frac{\partial v}{\partial y} \end{bmatrix} = \begin{bmatrix} 2y & 2x \\ -2x & 2y \end{bmatrix}$$

$dudv = |J(u,v)| \times dxdy \rightarrow \begin{bmatrix} 2y & 2x \\ 2x & -2y \end{bmatrix} \times dxdy = (4y^2 + 4x^2)dxdy$ The new region looks like The new region looks like this,

Therefore,

$$\iint (x^2+y^2)dA = \int_9^{16}\int_6^{14} (x^2+y^2)/(4y^2+4x^2)dudv = \int_9^{16}\int_6^{14} 1/4dudv = (16-9)\times(14-6)\times 1/4 = 14.$$

Example7:

Compute the following expression:

$$\iint \left(1/\sqrt{4x^2+9y^2}\right)dA$$

where dA is the region bounded by $(2x)^2+(3y)^2=36$.

Solution:

Again, let's graph the region.

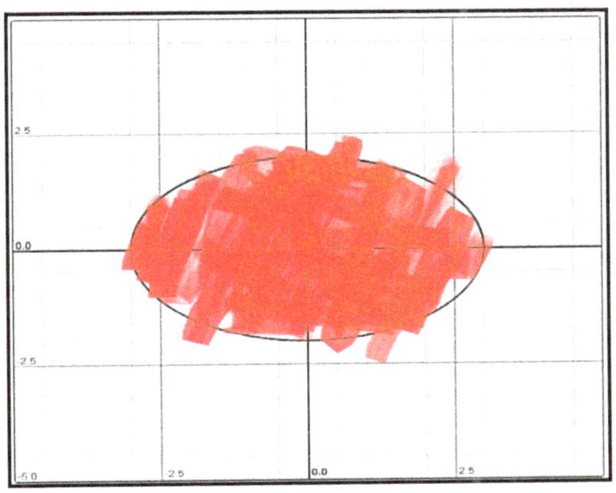

There are many ways to approach this, but to start, let's choose the change of variables to be $x=u/2,\ y=v/3$. Why?

$$(2x)^2+(3y)^2=36 \rightarrow u^2+v^2=36$$

The region now looks like this,

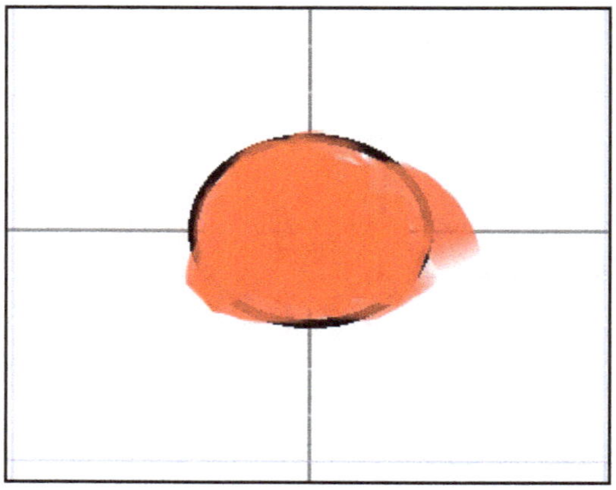

We've turned the region from an ellipse to a circle.

However, notice that in this case, instead of substituting u= and v=, we're substituting with x= and y=. Therefore the Jacobian will be written as J(x,y) and the expression will be dxdy=J(x,y)dudv.

$$x = u/2 \rightarrow \frac{\partial x}{\partial u} = 1/2, \frac{\partial x}{\partial v} = 0$$

$$y = v/3 \rightarrow \frac{\partial y}{\partial u} = 0, \frac{\partial y}{\partial v} = 1/3$$

$$J(x,y) = \begin{bmatrix} \frac{\partial x}{\partial u} & \frac{\partial x}{\partial v} \\ \frac{\partial y}{\partial u} & \frac{\partial y}{\partial v} \end{bmatrix} = \begin{bmatrix} 1/2 & 0 \\ 0 & 1/3 \end{bmatrix}$$

$$dxdy = |J(x,y)| \times dudv = \begin{bmatrix} 1/2 & 0 \\ 0 & 1/3 \end{bmatrix} \times dudv = (1/6)dudv$$

$\iint \left(1/\sqrt{4x^2+9y^2}\right)dA = \iint 1/6 * 1\sqrt{u^2+v^2}\ dudv$ Now, let's change the coordinates to polar:

$$rsin\theta = u \quad ,rcos\theta = v,$$

$$\iint 1/6 \times 1/\sqrt{u^2+v^2}\ dA = 1/6 \times \iint 1/\sqrt{r^2} \times r \times drd\theta = 1/6 \times \int_0^{2\pi}\int_0^6 drd\theta = 1/6 \times 36 \times \pi = 6\pi$$

Example 8:

Compute the following expression:

$\iint\left(\sqrt{(x-y)(x+6y)}\right)dA$ where dA refers to the plane bounded by the points (0,0), (1,1), (7,0), (6,-1).

Solution:

So far, we've only discussed cases where separation of variables is used to make regions easier to integrate over. For this example, separation of variables will be used to make the integrand easier.

First, let's graph the region.

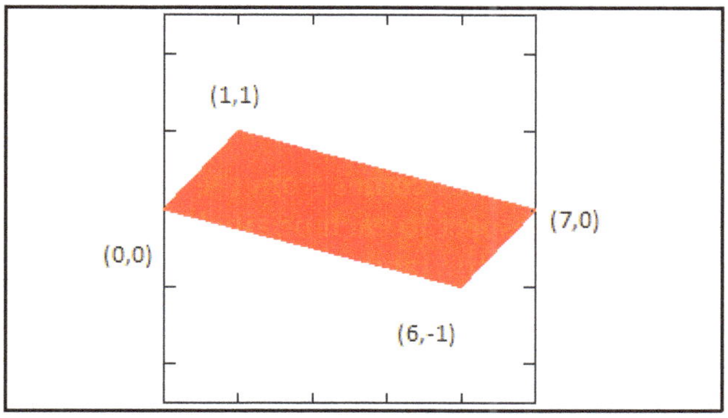

As previously stated, it would relatively simple to just do the double integral with this region. However, the integrand is the real problem. And so, to make things simpler, we'll set,

$$u = x - y \rightarrow \frac{\partial u}{\partial x} = 1, \frac{\partial u}{\partial y} = -1$$

$$v = x + 6y \rightarrow \frac{\partial v}{\partial x} = 1, \frac{\partial v}{\partial y} = 6$$

Notice that in this case, we're back to substituting from the u= and v= side.

$$J(u,v) = \begin{bmatrix} \frac{\partial u}{\partial x} & \frac{\partial u}{\partial y} \\ \frac{\partial v}{\partial x} & \frac{\partial v}{\partial y} \end{bmatrix} = \begin{bmatrix} 1 & -1 \\ 1 & 6 \end{bmatrix}$$

$$dudv = | J(u,v) | \times dxdy = \begin{bmatrix} 1 & -1 \\ 1 & 6 \end{bmatrix} \times dxdy = (7)dxdy$$

Let's look at the transformed region. It just so happens that it made a 7 x 7 square!

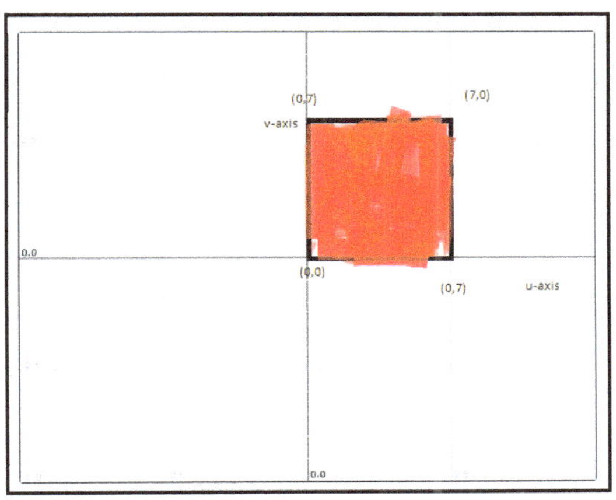

$$\iint \left(\sqrt{(x-y)(x+6y)}\right)dA = \int_0^7 \int_0^7 \sqrt{uv}dudv = 7*3/2*3/2 = 63/4$$

Transforming Tangent Vectors

The major application of Jacobian Matrices comes from the fact that it is made of the partial derivatives of the transformation with respect to each original element. Therefore, when a function F is transformed by a transformation T, the tangent vector of F at a point is likewise transformed by the Jacobian Matrix.

Here is a picture to illustrate this.

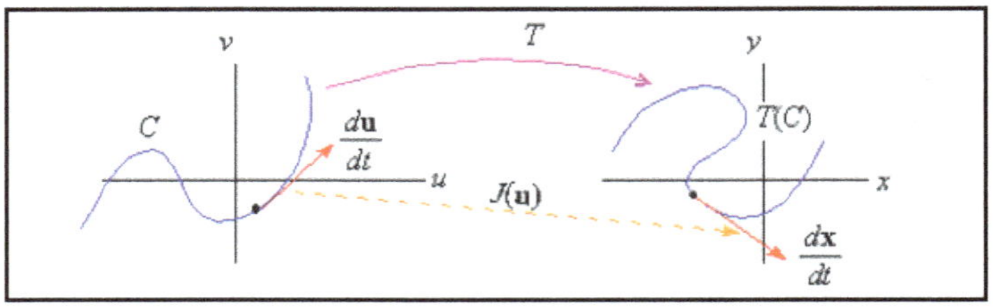

In other words, if we transform a function, we can find the new tangent vector at a transformed point.

Example:

$$<u,v>=<t^2,t^4>,\ T(u,v)=<u^3-v^3,3uv>=<x,y>$$

Find the velocity vector of $T(u,v)$ at $t=1$.

Solution:

$$x = u^3 - v^3 \rightarrow \frac{\partial x}{\partial u} = 3u^2, \frac{\partial x}{\partial v} = -3v^2$$

$$y = 3uv \rightarrow \frac{\partial y}{\partial u} = 3v, \frac{\partial y}{\partial v} = 3u$$

$$J(x,y) = \begin{bmatrix} \frac{\partial x}{\partial u} & \frac{\partial x}{\partial v} \\ \frac{\partial y}{\partial u} & \frac{\partial y}{\partial v} \end{bmatrix} = \begin{bmatrix} 3u^2 & -3v^2 \\ 3v & 3u \end{bmatrix}$$

$$u'(t) = \begin{bmatrix} 2t \\ 4t^3 \end{bmatrix}$$

$$J(x,y)\times u'(t)\begin{bmatrix} 3u^2 & -3v^2 \\ 3v & 3u \end{bmatrix} \times \begin{bmatrix} 2t \\ 4t^3 \end{bmatrix} = \begin{bmatrix} 3u^2\times 2t - 3v^2 \times 4t^3 \\ 3v \times 2t + 3u \times 4t^3 \end{bmatrix} = \begin{bmatrix} 3(t^2)^2 \times 2t - 3(t^4)^2 \times 4t^3 \\ 3\left(t^4\right) \times 2t + 3\left(t^2\right) \times 4t^3 \end{bmatrix} = \begin{bmatrix} 6t^5 - 12t^{11} \\ 18t^5 \end{bmatrix}$$

Therefore at t=1, the transformed tangent vector is:

$$\begin{bmatrix} -6 \\ 18 \end{bmatrix}$$

Linearization of Systems of Differential Equations

There is one last major application of Jacobians: approximating systems of Differential equations. Let's take the general expression:

$$\vec{x}' = f(\vec{x}, t)$$

Around a point, if $f(x,t)$ is differentiable, the system can be approximated to:

$$\vec{x}' \approx A\vec{x}, \quad A = J(\vec{x}, t_0)$$

The reason this works is similar to the idea behind one dimensional differentials:

$$dy \approx f'(x) \times dx$$

Why is this important? This can help us check stability for certain systems of differential equations. It is worth mentioning that this linearization works even when there are terms with t in them in the equation. This is because at very small intervals, those terms are small and don't affect the approximation.

Example 1:

Find the stability of the all the critical points of,

$$\begin{bmatrix} x' \\ y' \\ z' \end{bmatrix} = \begin{bmatrix} z - x \\ 3/(2+x) - 3y \\ 3y - z \end{bmatrix}$$

Solution:

To find the critical points, we set x'=y'=z'=0.

$$z = x, \quad y = 1/3x, \quad 3/(2+x) - x = 0 \to (x+3)(x-1) = 0, \quad (1,1/3,1) \quad (-3,-1,-3)$$

Now let's find the Jacobian at (1,1/3,1),

$$J(x', y', z') = \begin{bmatrix} -1 & 0 & 1 \\ -3/(2+x)^2 & -3 & 0 \\ 0 & 3 & -1 \end{bmatrix} = \begin{bmatrix} -1 & 0 & 1 \\ -1/3 & -3 & 0 \\ 0 & 3 & -1 \end{bmatrix}$$

The eigenvalues are -3.2055694304005904, -0.8972152847997048 ± 0.6654569511528134i.

Since the real parts are all negative, this critical point is locally stable.

Now let's find the Jacobian at (-3,-1,-3),

$$J(x',y',z')=\begin{bmatrix}-1 & 0 & 1\\ -3/(2+x)^2 & -3 & 0\\ 0 & 3 & -1\end{bmatrix}=\begin{bmatrix}-1 & 0 & 1\\ -3 & -3 & 0\\ 0 & 3 & -1\end{bmatrix}$$

The eigenvalues are -4, -2.5 ± 1.65831i. Because all of the real parts are negative, this critical point is locally stable.

Example 2:

Find the stability of the all the critical points of,

$$\begin{bmatrix}x'\\ y'\\ z'\end{bmatrix}=\begin{bmatrix}z+2y^2+x\\ y-z^{3/2}\\ sin(z)\end{bmatrix}$$

Solution:

The only critical point in this case is (0,0,0)

Now, let's do the Jacobian.

$$J(x',y',z')=\begin{bmatrix}1 & 2y & 1\\ 0 & 1 & 3/2*z^{1/2}\\ 0 & 0 & \cos(z)\end{bmatrix}=\begin{bmatrix}1 & 0 & 1\\ 0 & 1 & 0\\ 0 & 0 & 1\end{bmatrix}$$

The eigenvalue is just 1. Since this is positive, the critical point is unstable.

Example 3:

Find the stability of the all the critical points of,

$$\begin{bmatrix}x'\\ y'\end{bmatrix}=\begin{bmatrix}y\\ -9sin(x)-y/5\end{bmatrix}$$

Solution:

for critical points, we get that,

$$y=0,\quad x=\pi\times k,\quad k\in N$$

Now, let's do the Jacobian.

$$J(x',y') = \begin{bmatrix} 0 & 1 \\ cos(x) & -1/5 \end{bmatrix}$$

$$Case1): x = 2\pi k \rightarrow J(x',y') = \begin{bmatrix} 0 & 1 \\ 1 & -1/5 \end{bmatrix}$$

$$Case2): x = \pi + 2\pi k \rightarrow J(x',y') = \begin{bmatrix} 0 & 1 \\ 1 & -1/5 \end{bmatrix}$$

For Case 1, the eigenvalues are -1.104987562112089 and 0.904987562112089 . Therefore, those points are saddles.

For Case 2, the eigenvalues are -0.1-0.99498743710662î and -0.1+0.99498743710662î. Therefore those critical points are spiral sinks.

Since this system is 2-d, it is possible to give a phase portrait and show you that this linearized system is accurate.

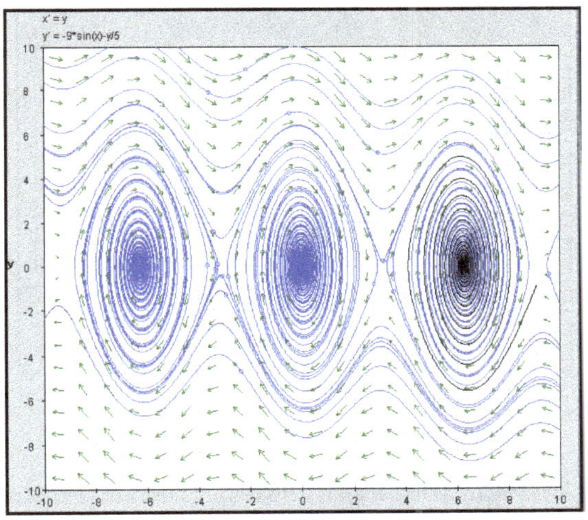

Example 4:

Find the stability of the all the critical points of,

$$\begin{bmatrix} x' \\ y' \end{bmatrix} = \begin{bmatrix} 2x - y - x^2 \\ x - 2y + y^2 \end{bmatrix}$$

Solution:

For critical points, we have (0,0) and (1,1). Now, let's do the Jacobian.

$$J(x',y') = \begin{bmatrix} 2-2x & -1 \\ 1 & -2+2y \end{bmatrix}$$

$$Case1): x = 0, \quad y = 0 \rightarrow J(x', y') = \begin{bmatrix} 2 & -1 \\ 1 & -2 \end{bmatrix}$$

$$Case2): x = 1, \quad y = 1 \rightarrow J(x', y') = \begin{bmatrix} 0 & 1 \\ 1 & 0 \end{bmatrix}$$

In case 1, we get eigenvalues of 1.732 and -1.732. So, that point is a saddle.

In case 2, we get eigenvalues of -i and i. This is important because this is an exception to the linearization precision so far. Eigenvalues of -i and i would tell us that the critical point is a center. However, for this case only, the linearization will not be precise, and the actual graph may not have this critical point as a center.

Here's a phase diagram to show this:

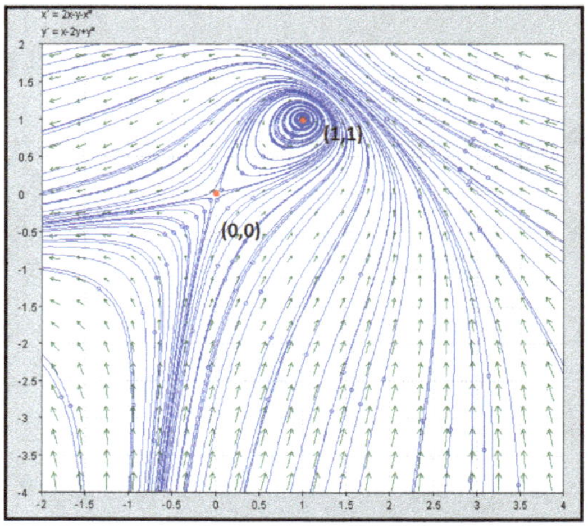

References

- Katz, Victor J. (1979-01-01). "The History of Stokes› Theorem". Mathematics Magazine. 52 (3): 146–156. doi:10.2307/2690275. JSTOR 2690275
- "Jacobian - Definition of Jacobian in English by Oxford Dictionaries". Oxford Dictionaries - English. Archived from the original on 1 December 2017. Retrieved 09, January 2020
- Theresa, M. Korn; Korn, Granino Arthur. Mathematical Handbook for Scientists and Engineers: Definitions, Theorems, and Formulas for Reference and Review. New York: Dover Publications. pp. 157–160. ISBN 0-486-41147-8
- "Jacobian - Definition of Jacobian in English by Oxford Dictionaries". Oxford Dictionaries - English. Archived from the original on 1 December 2017. Retrieved 25, June 2020
- Korn, Granino Arthur and Theresa M. Korn. Mathematical Handbook for Scientists and Engineers: Definitions, Theorems, and Formulas for Reference and Review. New York: Dover Publications. pp. 157–160. ISBN 0-486-41147-8

Permissions

All chapters in this book are published with permission under the Creative Commons Attribution Share Alike License or equivalent. Every chapter published in this book has been scrutinized by our experts. Their significance has been extensively debated. The topics covered herein carry significant information for a comprehensive understanding. They may even be implemented as practical applications or may be referred to as a beginning point for further studies.

We would like to thank the editorial team for lending their expertise to make the book truly unique. They have played a crucial role in the development of this book. Without their invaluable contributions this book wouldn't have been possible. They have made vital efforts to compile up to date information on the varied aspects of this subject to make this book a valuable addition to the collection of many professionals and students.

This book was conceptualized with the vision of imparting up-to-date and integrated information in this field. To ensure the same, a matchless editorial board was set up. Every individual on the board went through rigorous rounds of assessment to prove their worth. After which they invested a large part of their time researching and compiling the most relevant data for our readers.

The editorial board has been involved in producing this book since its inception. They have spent rigorous hours researching and exploring the diverse topics which have resulted in the successful publishing of this book. They have passed on their knowledge of decades through this book. To expedite this challenging task, the publisher supported the team at every step. A small team of assistant editors was also appointed to further simplify the editing procedure and attain best results for the readers.

Apart from the editorial board, the designing team has also invested a significant amount of their time in understanding the subject and creating the most relevant covers. They scrutinized every image to scout for the most suitable representation of the subject and create an appropriate cover for the book.

The publishing team has been an ardent support to the editorial, designing and production team. Their endless efforts to recruit the best for this project, has resulted in the accomplishment of this book. They are a veteran in the field of academics and their pool of knowledge is as vast as their experience in printing. Their expertise and guidance has proved useful at every step. Their uncompromising quality standards have made this book an exceptional effort. Their encouragement from time to time has been an inspiration for everyone.

The publisher and the editorial board hope that this book will prove to be a valuable piece of knowledge for students, practitioners and scholars across the globe.

Index

www.ingramcontent.com/pod-product-compliance
Lightning Source LLC
LaVergne TN
LVHW080908210926
848485LV00004B/615